# Clinical Ethics for Consultation Practice

Joseph T. Bertino

# Clinical Ethics
# for Consultation Practice

 Springer

Joseph T. Bertino
UPMC Health System
Pittsburgh, PA, USA

ISBN 978-3-030-90181-3    ISBN 978-3-030-90182-0   (eBook)
https://doi.org/10.1007/978-3-030-90182-0

This Springer imprint is published by the registered company Springer Nature Switzerland AG
The registered company address is: Gewerbestrasse 11, 6330 Cham, Switzerland

*For my family*

# Foreword

As the first quarter of the twenty-first century nears an end, those bioethicists who work primarily as individual clinical ethics consultants in American academic medical centers, healthcare systems, and community hospitals may feel a bit disheartened. For the past several years, many have hoped that the movement toward greater "professionalization" of the "field" would be making more headway.[1]

After decades of conversation, the American Society for Bioethics and Humanities (ASBH) instituted the Healthcare Ethics Consultant Certification (HEC-C) program in 2018 [1].[2] Regrettably, there appears to be no significant or measurable impact to date. Still, it may be too early to evaluate the effort appropriately [2]. There may be several reasons why this "certified professional" status is not gaining broader general recognition [3]. Persons certified by the ASBH-sponsored HCEC Certification Commission are "healthcare ethics consultants," not clinical ethics consultants [4]. There is a difference. Clinical ethics consultants are a subset of healthcare ethics consultants. The work is more narrowly defined and highly patient focused [5]. Many of those who are identified by the HEC-C credential indeed provide individual clinical ethics consultation services. However, it is also true that some certified individuals do not labor in clinical settings very often, if at all. For some, the single credential designation, crafted for a wide spectrum of "practitioners" with "clinical ethics experience," is confusing, particularly since an

---

[1] For discussion purposes, one might define an *individual clinical ethics consultant* as one who can offer clinical ethics consultation services without any assistance from others (i.e.,, singularly, alone, not as a member of a small team or large committee) as contemplated by the ASBH's *Core Competencies for Healthcare Ethics Consultation*, 2d ed. Chicago, Illinois: American Society for Bioethics and Humanities, 2013. See also, Tarzian, Anita J., & ASBH Core Competencies Update Task Force: Health Care Ethics Consultation: An Update on Core Competencies and Emerging Standards from the American Society for Bioethics and Humanities' Core Competencies Update Task Force. Am J Bioeth. 2013;13(2):3-13.

[2] One of the first articles to suggest uniform practice standards (a precursor to certification) was published in 1994. Fletcher JC, Hoffmann DE. Ethics Committees: Time to Experiment With Standards. Ann Intern Med. 1994 (Feb 15);120(4):335-8.

individual with the HEC-C label may or may not be a clinical ethics consultant, let alone an individual clinical ethics consultant [6].

Moreover, some of those with the HEC-C credential may leave the lesser informed with the impression that they are qualified and can – as an individual – provide clinical ethics consultation services, even though they lack the requisite essential competencies and skills for that level of involvement as differentiated and endorsed by the ASBH.[3] The HEC-C Web site may contribute to the clutter. In reviewing the "About HEC-C" information page, one can see the potential for mis-information. Per the Commission, *healthcare ethics consultation*: "a set of services provided by *an individual or group* in response to questions from patients, families, surrogates, healthcare professionals, or other involved parties who seek to resolve uncertainty or conflict regarding value-laden concerns that emerge in health care [emphasis added]." [7] The definition clearly emphasizes bedside patient care uncertainties or conflicts – that is, clinical ethics – but one who has never been involved in any way with patients or families or providers (e.g., "those in leadership roles within their institutions' healthcare ethics initiatives" or college instructors who teach medical ethics topics to undergraduates) may qualify to sit for the HEC-C certification examination. The HCEC Commission – like most other professions' credentialing bodies – does not police the practices of certified practitioners. There are two requirements to qualify for the written HEC-C certifying examination: (1) a "minimum of a Bachelor's Degree" and (2) "400 hours of clinical ethics experience, related to the major domains of the content outline, within the previous [four] years." The application process requires potential candidates to self-attest that they meet the examination qualifications.

To illustrate how nominal the "experience" requirements for examination are, one might cite an example. An individual who holds a liberal arts baccalaureate degree, serves monthly as a "community member" on an institutional ethics com-mittee for an hour each meeting for 4 years, and "attends" a 2-day conference or workshop annually for those same 4 years period would probably meet the eligibil-ity requirements for the examination. As a second illustration, one might use the same set of facts for a person who serves a similar amount of time as a member of an institutional review board (IRB) which deals exclusively with healthcare research ethics issues. Again, at first blush, this person too meets the examination eligibility requirements. At the other end of the qualifications spectrum – an individual who holds a professional or graduate degree and a state's license to practice in a health-care field, has completed a 1 to 2-year fellowship in clinical ethics, and who is involved in over 1000 hours or more annually in providing individual clinical ethics consultation services (while often concurrently engaging in related educational and research activities) qualifies to sit for the examination.

---

[3] Information about the ASBH educational core competencies may be found at: http://www.asbh.org/uploads/files/pubs/CCUpdateNov09.pdf (accessed August 10, 2021). A companion mono-graph to the ASBH *Core Competencies for Health Care Ethics Consultation* is the ASBH *Improving Competencies in Clinical Ethics Consultation: An Education Guide.* Information about the *Guide* may be found at: http://www.asbh.org/publications/content/edguide.html.

The more highly trained and mentored clinical ethics consultant will probably view the examination requirements for those who barely qualify as simply too low to warrant comparison. Moreover, an online examination of 110 multiple choice questions (to be completed in three hours) with an overall 98.6% pass rate may not be enough to establish that one is "a leader in the evolution of healthcare consulting" and fails to "underscore … skills with a national standard" towards "[maintaining] the highest quality of patient and family care." [8] The *de minimis* standard for HEC-C certification is much less than what numerous nationally recognized individual clinical ethics consultants had envisioned during a heated 5-year-long literature debate that led up to the creation of the ASBH-backed certification program [9–12]. For any certification to be meaningful, the accrediting body should not establish examination criteria based on a survey of pre-candidates preferences about educational background and examination costs [13]. A more cynical observer might question the value of a hallow credential or empty title established by a trade association to placate a fraction of its multidisciplinary and diverse membership base while possibly creating a revenue stream that contributes to improved financial stability for the organization because alternatives were just too complicated or expensive [14]. An even more crass view might involve creating multiple complimentary certification programs, perhaps a Clinical Ethics Consultant – Certified (CEC-C) designation, as other trade associations have done.[4] Low certification standards reflect poorly on all practitioners when they underpin the philosophy that an amateurish approach to providing clinical ethics services by poorly prepared persons or teams is an acceptable national norm.

However, apart from any – what might be perceived as – internecine professional rivalry or cultural war about credentialing, individual clinical ethics consultants might be alarmed by recent study findings reported by Ellen Fox and colleagues [15, 16]. The 2021 report is a follow-up done about 15 years after a similar survey of ethics consultation programs in US hospitals also conducted by Fox and colleagues [17]. In the 2007 survey, the 591 respondent hospitals (from a randomly selected sample of 600/5072 stratified by bed size) reported a median of 3 annual consults for all hospitals and 3 annual consults for those hospitals with 400 beds or more. In the 2021 study, the 462 respondent general hospitals (again from a randomly selected sample of 600/~5534 stratified by bed size) reported a median of 3 annual consults for all hospitals and 6 annual consults for those hospitals with 400 beds or more. (Hospitals with 400 beds or more accounted for only about 8% of hospitals in 2016 [18].) Fox described the comparison as a "wake-up call" for leading bioethicists [19]. Naturally, clinical ethics consultation practices have changed

---

[4] For example, the Health Care Compliance Association sponsors a Compliance Certification Board that offers four individual certifications as well as corporate certification. See: https://www.hcca-info.org/certification/become-certified (accessed August 16, 2021). The certifications qualifications include work experience, earned continuing education credits (offered by the association), and an examination. Each examination costs $275 for members and $375 for non-members. The CCB time-sensitive certifications may be renewed with application, proof of continuing education, and payment of a $115 fee for members or $225 for non-members.

significantly over the past 30-odd years [20]. Also, one might suspect that significant advances and innovations have occurred within academic medical centers (only about 5% of US hospitals), and therefore had no statistical impact on the overall results. Moreover, the 2021 survey found no change in the average percentage of ethics consultation practitioners who had completed an advanced degree or a fellowship (about 8% in both surveys). Additionally, the 2021 national survey results mirror some state surveys completed between the two national Fox et al. studies [21].

In assessing the latest national survey results, one should be cautious about attempting apple-to-apple comparisons. It may not be possible: only a few dramatic reasons are necessary to illustrate the limitations. First, virtually all medical schools and residency programs now teach students and physicians-in-training about clinical ethics dilemmas [22]. Such was not the case 30 years ago [23]. Today, several medical schools and residency and fellowship programs use clinical ethics simulations as a routine exercise to help prepare for real-life situations [24]. Second, faculty physicians and hospitalists in most larger hospitals have replaced community primary care providers as the attendings-of-record for most admitted patients [25]. In-house patient care teams – who were strangers to patients and their families before admission – are routine. Some teams now include clinical pharmacists and clinical ethicists on bedside rounds, particularly in intensive care units [26, 27]. Separate consultation requests are less likely when the individual clinical ethics consultant participates in the day-to-day medical decision-making processes. Third, palliative care programs and palliative care consultation services are widely available today in almost every US hospital, regardless of size [28]. Since many of the earlier ethics consultations dealt with "death and dying" concerns, there is less need for an ethics consult when palliative care providers are uniquely positioned to offer advice about pain and symptom management, provide patient and family support, and counsel about advance planning issues.

Moreover, because Medicare and most insurance plans pay for palliative care and hospice services, hospitals and systems have identified an ongoing payment mechanism for ethics-related costs that were previously part of the workload of salaried clinical ethicists [29]. Fourth, institutional ethics policies and practices have matured over the years. Some hospitals have instituted protocols that help identify and resolve potential clinical ethics problems (such as identifying the patient's legally authorized representative on admission) long before any concerns can reach the point of disagreement [30]. Also, today, the overwhelming majority of hospitals – which were reluctant to allow ethics consultations without an order from the attending physician – have eliminated consultation barriers and now allow any interested party to request a consultation [31]. And fifth, in responding to various lobbying pressures and personal requests, most state legislatures have enacted statutes creating uniform advance healthcare planning documents and decision-making standards, such as improved durable powers of attorney for healthcare and Physician's Orders for Life-Sustaining Treatment (POLST) forms and default proxy processes for patients who lack capacity [32]. In sum, healthcare delivery concerns for clinical ethics consultants in 2007 and 2021 are exceedingly difficult because the landscapes are too different.

In addition to Fox et al. surveying hospitals about the structure, organization, activities, and funding of their healthcare ethics programs, it might have been helpful to look closely at the dilemmas clinical ethics consultants are asked to ponder and perhaps resolve. As far as valuable comparisons between 2007 and 2021, it would be interesting to know if and how the clinical ethics involved in the consultation requests have changed. The follow-up survey confirms that the consultation requests have nearly doubled numerically. Still, there is little information about the questions posited. How the questions asked by those requesting consultations are different – if at all – may be very relevant in assessing practice changes [33]. Previously, the critical issues resulting in clinical ethicists' involvement were relatively predictable: Who is the patient's legally authorized representative when the patient lacks capacity? How best to handle what appears to be irresolvable conflicts between family members? How might the team proceed when the medical uncertainties are overwhelming and frustrating, and the care interventions add to the patient's suffering? [34] Because these issues and answers are not the exclusive domain of clinical ethicists (and because we can assume that clinical ethicists are not involved in all hospital cases in which these concerns arise), thought leaders understand that other professionals besides clinical ethicists participate in dilemma discussion and resolution. For example, the hospital's lawyer may identify the appropriate representative for patients who lack capacity; social workers, among others, help resolve family conflicts; and pastoral care workers, medical consultants, and trusted colleagues often counsel bedside caregivers in dealing with moral distress. Individual clinical ethics consultants may be involved solely. Still, more often, they supplement and strengthen existing dilemma resolution networks and mechanisms. In some cases, they reassure the team by endorsing a plan of action all agree is "the right thing to do in the patient's best interests." [35]

However, it may be unwise to paint too bleak an assessment about the future of professionalization at this juncture. All together – notwithstanding the raucous credentialing debate and the 2021 Fox et al. survey showing what appears to be the clinical ethics movement hitting a brick wall – there are silver linings. The present clinical ethics professionalization movement may be better characterized as a glass half full rather than half empty. For those individual clinical ethics consultants that practice in large facilities and systems and academic medical centers, the HEC-C credential is quite possibly irrelevant, at least in the near term. It is unlikely that supervisors hiring clinical ethicists at the nation's largest 15% to 20% of hospitals by bed size will employ clinical ethicists for open positions who are minimally qualified. The HEC-C label may matter at some facilities where most of the country's clinical ethics consults are performed, but not in the same way as at those institutions that employ a robust or larger staff of clinical ethicists. Suppose the HEC-C is required of new hires at these large hospitals. In that case, it is probably more to show solidarity with the ASBH than to meet an aspirational quality standard. Moreover, the increasing consultation volume – as poor an indicator as this may be for demonstrating superior clinical ethics consultation services delivery – at the larger hospitals is growing, according to the 2021 national survey. Clearly, those who request consultations must continue to ask for assistance because of some practical – or beneficial – reason. Moreover, clinical ethicists at large hospitals are valued for their clinical ethics consultation efforts and their local services. For example, they might directly support institutional ethics committees, assist appropriately in those cases with an extended length of stays, and offer educational opportunities for staff and employees at all levels and in various settings, and scholarly activities such as theoretical and empirical research [36–41]. Moreover, one would hope that future clinical ethicists might better sort out expertise within institutions and avoid unnecessary turf battles or competitions with complimentary but crucial departments like finance, legal, compliance, research ethics, and palliative care.

More critical to the professionalization of clinical ethics are those who offer bedside consultations in the nation's hospitals and medical centers. Individuals like Joseph Bertino and others similarly situated who working in the "trenches," advancing the field by carrying on their day-to-day activities.[5] He and his peers are the heirs of America's "pioneer" clinical ethicists [42]. At the national level, it will be for today's clinical ethicists to carry on the work by developing more uniform and standardized processes for reporting consultations among institutions, crafting curricula future clinical ethics consultants and accrediting training programs, and writing [43–45]. Perhaps the time has arrived when more practicing individual clinical ethics consultants should be producing the next generation of texts. Bertino's *Clinical Ethics for Consultation Practice* – which uses an Aristotelian-Pellegrino framework to explore integrating medical ethics with normative theory – may be one example. It is time for serious clinical ethics scholars to move past books about processes and troublesome cases to revisit their roots [46–48]. Like the works written by some of the pioneer clinical ethics scholars 50 years ago, Bertino's book adds to the list of possible aids for those who participate in few consultations per year and those who train and mentor others who will one day fill future full-time clinical ethics positions [49]. Practical clinical ethics – along with medicine and the healing arts – rests on a foundation of moral philosophy that is just as important to the student clinical ethicist as is law, public policy, justice and equity, and research. It is a critical task for educators to help those who follow.

---

[5] "Trenches" was a word used by Albert R. Jonson – one of the founding fathers of American bioethics – in an oft-told allegory featuring balloons, bicycles, physicians, and ethics pundits. Jonson observed that it was difficult for one riding in a balloon way overhead to help a bicyclist, peddling feverishly far below, traverse rugged terrain. The balloonist's view was at the 5000-feet-level and of little help. He suspected that the bicyclist was reasonable to ignore the balloonist's proffered advice, even if he could hear it from a distance. Though knowledgeable about much, the balloonist could not understand the lay of the land because he was not in the trenches with the bicyclist. For Jonson, the bicyclist represented the overwhelmed physician dealing with ethical dilemmas at the patient's bedside, and the balloonist was the ethics expert trying to help the physician navigate dilemmas. Andereck W. The Balloon, the Bicycle, and Al Jonson. The Hastings Center. 2020: Nov 4. Available at: https://www.thehastingscenter.org/the-balloon-the-bicycle-and-al-jonsen/ (accessed August 15, 2021).

It is reasonable for those blazing new professional trails to debate future directions and paths. One recognizes that this is the usual course for evolving professions or fields; a recent example is a continuing growth of "clinical pharmacy." For that reason, and because the hope for the better is so bright, many individual clinical ethicists trust that the conversation about professionalization shifts soon to where the focus, away from credentials and numbers, is towards the quality of services provided and the scholarship produced [50].

Bruce White
University of Tennessee College of Medicine
Nashville, TN, USA

# References

1. Hynds JA, Raho JA. A Profession Without Expertise? Professionalization in Reverse. Am J Bioeth. 2020;20(3):44-46.
2. Horner C, Childress A, Fantus S, Malek J. What the HEC-C? An Analysis of the Healthcare Ethics Consultant-Certified Program: One Year in. Am J Bioeth. 2020 Mar;20(3):9-18.
3. Siegler M. The ASBH Approach to Certify Clinical Ethicists Is Both Premature and Inadequate. J Clin Ethics. 2019;30(2):109-116.
4. Healthcare Ethics Consultant Certification. American Society for Bioethics and Humanities. 2021. Available at: https://heccertification.org/about/certification-commission (accessed August 10, 2021).
5. La Puma J, Stocking CB, Silverstein MD, DiMartini A, Siegler M. An Ethics Consultation Service in a Teaching Hospital. Utilization and Evaluation. JAMA. 1988 Aug 12;260(6):808-11.
6. *Healthcare Ethics Consultant Certified (HEC-C) Examination Candidate Handbook*. Chicago, Illinois: Healthcare Ethics Consultation Commission, 2018 (June 1). https://asbh.org/uploads/certification/HEC-C_Candidate_Handbook.pdf (accessed August 13, 2021).
7. Healthcare Ethics Consultant-Certified Program. Health Care Ethics Certification Commission. 2021. Available at: https://asbh.org/certification/hcec-certification (accessed August 13, 2021).
8. Healthcare Ethics Consultant-Certified Program. Health Care Ethics Certification Commission. 2021. Available at: https://heccertification.org (accessed August 13, 2021).
9. Dubler NN, Blustein J. Credentialing Ethics Consultants: An Invitation to Collaboration. Am J Bioeth. 2007 Feb;7(2):35-7.
10. Dubler NN, Blustein J. Credentialing Ethics Consultants: An Invitation to Collaboration. Am J Bioeth. 2007 Feb;7(2):35-7.
11. Kodish E, Fins JJ, Braddock C 3rd, et al. Quality Attestation for Clinical Ethics Consultants: A Two-Step Model From the American Society for Bioethics and Humanities. Hastings Cent Rep. 2013 Sep-Oct;43(5):26-36.
12. White BD, Jankowski JB, Shelton WN. Structuring a Written Examination to Assess ASBH Health Care Ethics Consultation Core Knowledge Competencies. Am J Bioeth. 2014;14(1):5-17.
13. Antommaria AHM, Feudtner C, Benner MB, Cohn F, & on Behalf of the Healthcare Ethics Consultant Certification Commission. The Healthcare Ethics Consultant-Certified Program: Fair, Feasible, and Defensible, But Neither Definitive Nor Finished. Am J Bioeth. 2020;20(3): 1-5.
14. Fins JJ, Kodish R, Cohn F et al. A Pilot Evaluation of Portfolios for Quality Attestation of Clinical Ethics Consultants, Am J Bioeth. 2016;16(3):15-24.

15. Fox E, Danis M, Tarzian AJ, Duke CC. Ethics Consultation in U.S. Hospitals: A National Follow-Up Study. Amer J Bioeth. 2021; Mar 26 (published online).
16. Danis M, Fox E, Tarzian A et al. Health Care Ethics Programs in U.S. Hospitals: Results from a National Survey. BMC Med Ethics. 2021; 22:107 (July 29) (open access 14 pp.).
17. Fox E, Myers S, Pearlman RA. Ethics consultation in United States hospitals: a national survey. Am J Bioeth. 2007 Feb;7(2):13-25.
18. Facts on U.S. Hospitals. American Hospital Association. 2021. Available at: https://www.aha.org/statistics/fast-facts-us-hospitals (accessed August 10, 2021).
19. Kusterbeck S. Survey: Ethics Consultations Gap Widening Between Small and Large Hospitals. Medical Ethics Advisor. 2021: Aug 1. Available at: https://www.reliasmedia.com/articles/148340-survey-ethics-consultations-gap-widening-between-small-and-large-hospitals (accessed August 10, 2021).
20. Farroni JS. Tumilty E, Mukherjee D, et al. Emerging Roles of Clinical Ethicists. J Clin Ethics. 2019;30(3):262-269.
21. Hughes R. Balancing Act – Ethics in Health Care: The Arizona Landscape. Arizona Health Futures. 2014: Aug. Available at: http://vitalysthealth.org/wp-content/uploads/2014/02/ib-08summer.pdf (accessed September 10, 2021).
22. Giubilini A, Milnes S, Savulescu J. The Medical Ethics Curriculum in Medical Schools: Present and Future. J Clin Ethics. 2016 summer; 27(2):129-45.
23. McElhinney TK, Pellegrino ED. The Institute on Human Values in Medicine: Its Role and Influence in the Conception and Evolution of Bioethics. Theoretical Medicine and Bioethics. 2001;22 (4):291-317.
24. Collins S, Crowther N, McCabe A et al. 0120 Simethics: Teaching medical ethics through high-fidelity simulation. BMJ Simulation and Technology Enhanced Learning. 2015; 1.
25. Pham HH, Devers KJ, Kuo S, Berenson R. Health Care Market Trends and the Evolution of Hospitalist Use and Roles. J Gen Intern Med. 2005 Feb;20(2):101-7.
26. Kucukarslan SN, Peters M, Mlynarek M, Nafziger DA. Pharmacists on Rounding Teams Reduce Preventable Adverse Drug Events in Hospital General Medicine Units. Arch Intern Med. 2003;163(17):2014–2018.
27. Vig EK. Weekly Rounding With the MICU Team: Description of a Clinical Ethics Project. Am J Hosp Palliat Care. 2019 Apr;36(4):290-293.
28. Hughes MT, Smith TJ. The Growth of Palliative Care in the United States. Annu Rev Public Health. 2014; 35:459-75.
29. Sulllender RT, Selenich SA. Financial Considerations of Hospital-Based Palliative Care. Research Triangle Park, North Carolina: RTI Press, 2016.
30. Hatler CW, Grove C, Strickland S, Barron S, White BD. The Effect of Completing a Surrogacy Information and Decision-Making Tool Upon Admission to an Intensive Care Unit on Length of Stay and Charges. J Clin Ethics. 2012 Summer;23(2):129-38.
31. Cederquist L, LaBuzetta JN, Cachay E et al. Identifying Disincentives to Ethics Consultation Requests Among Physicians, Advance Practice Providers, and Nurses: A Quality Improvement All Staff Survey at a Tertiary Academic Medical Center. BMC Med Ethics. 2021 Apr 23;22(1):44.
32. Sabatino CP. The evolution of health care advance planning law and policy. Milbank Q. 2010;88(2):211-239.
33. Bishop JP, Fanning JB, Bliton MJ. Of Goals and Goods and Floundering About: A Dissensus Report on Clinical Ethics Consultation. HEC Forum. 2009 Sep;21(3):275-91.
34. White BD, Zaner RM, Bliton M et al. An Account of the Usefulness of a Pilot Clinical Ethics Program at a Community Hospital. QRB/Quality Review Bulletin. 1993; 19(1):17-24.
35. Tapper EB. Consults for Conflict: The History of Ethics Consultation. Proc (Bayl Univ Med Cent). 2013;26(4):417-422.
36. White BD. Clinical Ethics Training for Members of New York Ethics Review Committees (ERCs). New York State Bar Association Health Law Journal. 2013;18(2):28-35.

37. Allen J. Reducing the Hospital Length of Stay. Hospital Medical Director. 2019 (Feb 11). Available at: https://hospitalmedicaldirector.com/reducing-the-hospital-length-of-stay/ (accessed August 16, 2021).
38. White BD, Zaner RM. Clinical Ethics Training for Staff Physicians: Results of a Pilot Project in a Community Hospital. J Clin Ethics. 1993; 4:229-235.
39. Nilstun T, Cuttini M, Saracci R. Teaching Medical Ethics to Experienced Staff: Participants, Teachers and Method. J Med Ethics. 2001;27(6):409-412.
40. Singer PA, Siegler M, Pellegrino ED. Research in Clinical Ethics. J Clinical Ethics. 1990 Summer;1(2):95-98-1.
41. Singer PA, Pellegrino ED, Siegler M. Clinical Ethics Revisited. BMC Med Ethics. 2001; 2:1.
42. White BD, Shelton WN, Rivais CJ. Were the "Pioneer" Clinical Ethics Consultants "Outsiders"? For Them, Was "Critical Distance" That Critical? Am J Bioeth. 2018 Jun;18(6):34-44.
43. Kornfeld DS, Prager K. The Clinician as Clinical Ethics Consultant: An Empirical Study. J Clin Ethics. 2019;30(2):96-108.
44. Guerin RM, Diekema DS, Hizlan S, Weise KL. Do Clinical Ethics Fellowships Prepare Trainees For Their First Jobs? A National Survey of Former Clinical Ethics Fellows. J Clin Ethics. 2020;31:372-380.
45. Shelton WN, White BD. The Process to Accredit Clinical Ethics Fellowship Programs Should Start Now. Am J Bioeth. 2016;16(3):28-30.
46. Guerin RM, Diekema DS, Hizlan S, Weise KL. Do Clinical Ethics Fellowships Prepare Trainees For Their First Jobs? A National Survey of Former Clinical Ethics Fellows. J Clin Ethics. 2020;31:372-380.
47. La Puma J, Schiedermayer D. *Ethics Consultation: A Practical Guide*. Boston, Massachusetts: Jones & Bartlett, 1994.
48. Ford PJ, Dudzinski DM. *Complex Ethics Consultations: Cases That Haunt Us*. Cambridge, UK: Cambridge University Press, 2008.
49. Jonsen A, Siegler M, Winslade W. *Clinical Ethics: A Practical Approach to Ethical Decisions in Clinical Medicine,* 9th ed. New York: McGraw Hill, 2021.
50. American College of Clinical Pharmacy Board of Regents, Maddux MS. Board of Regents Commentary. Qualifications of Pharmacists Who Provide Direct Patient Care: Perspectives on the Need For Residency Training and Board Certification. Pharmacotherapy. 2013 Aug;33(8):888-91.

# Preface

Clinical ethics is a growing field, though still in its early stages compared to the history of Western medical practice. Disciplines and subdisciplines within medicine have developed standards of practice, recognized accreditation systems, and licensures that justify the practice in the United States. In contrast, clinical ethics has not received the same kind of attention regarding professional recognition and the overall legitimacy of the field. This book examines clinical ethics as a vocational discipline and the educational methodologies used to train prospective ethicists. Though the state of clinical ethics has improved over the past two decades, tremendous work is still required if the field intends to adapt and grow alongside healthcare.

This project began as a doctoral dissertation in 2015, where I sought to wed philosophical concepts with practical instances in medicine. Naturally, I am not the first scholar to attempt this endeavor. Edmund Pellegrino and others have accomplished this task many times over. Still, I felt the field needed a new perspective, especially considering the exponential growth of clinical ethics as a vocational discipline and the influx of contemporary scholars in bioethics and medicine. More young professionals across various disciplines are not merely looking to pursue a challenge in clinical ethics nor seek a higher philosophical understanding of moral behavior. Instead, they are looking for employment. It is no surprise that philosophers, among other liberal arts scholars, struggle to find work in today's job market. Today's applicant pool for tenured track philosophy appointments is wildly competitive compared to the golden age of academic appointments in the 1950s and 1960s.

In contrast, the demand for clinical ethics consultants and healthcare ethics educators is growing every day. Nationwide recognition of the importance of educated moral experts in healthcare has grown exponentially. However, the motivation to establish clinical ethics consultation services is unclear. Moral theorists and bioethicists hope hospital systems seek to develop formidable clinical ethics consultation services for noble reasons. Still, investing resources into clinical ethics consultation services for many healthcare systems may only serve to promote an institution's image. Whatever the case, the growing demand for clinical ethics consultants presents an opportunity to develop a culture of ethical behavior.

My attempt to wed philosophical concepts with clinical ethics practice is admittedly lofty and, at times, perhaps unapologetically complex. While clinical ethics training is ripe with normative ethical theories like utilitarianism and deontological thought, few have examined the phenomenological impact of humans' relationship with modern technology and its moral implications. Genetic research and modification may be the most prominent venue for exploring the ethical implications of medicine and technology. However, the underlying theme of this discussion lies in our ability to respond to clinical situations in a timely and adaptive manner.

The American Society of Bioethics and Humanities (ASBH) has served as the primary organizational body in clinical ethics. Additionally, the ASBH has asserted itself as the field's leading accreditation body for clinical ethicists. The materials produced by the ASBH include practical and process skills for budding ethicists, which provide excellent information for those looking to practice clinical ethics in a structured manner. In 2018, the ASBH developed the only existing credential for clinical ethicists, the Healthcare Ethics Consultant Certified (HEC-C). This credential has granted various individuals a widely recognized certification title that provides legitimacy to the discipline. However, there exist different knowledge gaps in training clinical ethics that this book addresses. Specifically, this book advocates for a rigorous philosophical training methodology in normative ethics. By training prospective ethicists in normative ethics theories in conjunction with the established clinical ethics training methods, consultants can use analytic moral reasoning skills in vocational practice.

Various training methods and mechanisms for learning healthcare ethics provided me with a good foundation for practical application in my professional training. Other mandatory steps in my training revealed gaps in clinical ethics consultation training that became apparent only after my career began. Still, my formal education in moral philosophy and practical experience from my first position as a clinical ethicist have been the most beneficial steps in my training. Ultimately, this book is designed for philosophers and healthcare practitioners alike. One cannot simply jump into clinical ethics with only a base understanding of the scope and purpose of the practice. Instead, individuals must establish a foundational understanding of virtue and normative ethics and wed these concepts with practical consultation methods.

Whether you are a member of a hospital ethics committee, healthcare worker, or a philosopher seeking practical application of their skills, *Clinical Ethics for Consultation Practice* is a guide for improving one's moral reasoning and analytical thinking. Consistent practice honing one's analytic moral reasoning skills will inevitably provide budding moral theorists with a more refined standard of practice and further legitimize the field of clinical ethics.

Pittsburgh, PA, USA                                                                          Joseph T. Bertino

# Definitions

| | |
|---|---|
| **ACGME** | Accreditation Council for Graduate Medical Education. The main accreditation body for graduate physicians across the United States. |
| **ACPE** | Accreditation Council for Pharmacy Education. |
| **AND** | Allow Natural Death. A variably used code status determination in many circles. Though commonly misused and misunderstood, AND is synonymous with **DNR/DNI**. |
| **ASBH** | American Society of Bioethics and Humanities. The leading bioethics licensure and credentialing body in the United States. |
| **Cardiac Arrest** | The cessation of heart function. |
| **CASES** | A clinical ethics case methodology consisting of five components: Collect, Assess, Survey, Examine, Synthesize. |
| **CECA** | Clinical Ethics Consultation Affairs. A committee assembled in 2009 by the ASBH, tasked with designing and implementing credentialing standards. |
| **CMO** | Comfort Measures Only. A formal order wherein care teams maximize patients' comfort rather than pursue therapeutic or curative treatment. In these settings, medical teams typically do not introduce new interventions or escalate existing interventions. |
| **Code Status** | The established resuscitation plan for patients in the event of cardiac arrest. |
| **Common Morality** | A moral methodology and ethical intention agreed upon by a community of individuals—moral reasoning by a community of professionals who invariably seek to resolve moral discrepancies. |
| **CPE** | Clinical Pastoral Education. A graduate medical education for existing hospital chaplains. |
| **CRRT** | Continuous Renal Replacement Therapy. A medical therapy that replaces the function of the kidneys. |

| | |
|---|---|
| **DNR/DNI** | Do not resuscitate/Do not intubate. A code status indicating to care teams to not provide resuscitative efforts nor intubation in the event of respiratory or cardiac arrest. |
| **ECMO** | Extracorporeal Membrane Oxygenation. A medical intervention wherein blood is mechanically oxygenated outside of the body. |
| **EHBP** | European Hospital-Based Bioethics Program. A formal educational assessment and development program designed to refine and improve European hospital-based bioethics. |
| **EMR** | Electronic Medical Record. The modern standardization of medical documentation. |
| **Four Topics** | Also referred to as "Four box method" or "Four quadrants," is an ethics consultation methodology, developed by AR Jonsen, M Siegler, W Winslade, that examines the clinical indiciations, patient preferences, quality of life, and contextual features of a patient. |
| **HCEC** | Health Care Ethics Consultation. A service that aims to facilitate discussion and develop resolution strategies between stakeholders who have difficulty resolving a value-laden issue by bolstering communication between healthcare professionals, patients, surrogates, and team members. |
| **HGP** | Human Genome Project. A scientific research team established under the Clinton administration, tasked with identifying and mapping the genomic composition of human beings. |
| **JCAHO** | Joint Commission on Accreditation of Healthcare Organizations. An organized authority in American healthcare tasked with assessing and upholding a standard of care, safety, and operations for healthcare organizations. |
| **NPO** | Nothing by Mouth. From the Latin, *nil per* os, a physician's short-hand order for no nutrition or medications by mouth for individual patients, typically used before surgical interventions or to protect a patient's airway. |
| **NACC** | The National Association of Catholic Chaplains. A national accreditation body for hospital chaplains in the United States. |
| **Normative [Ethics]** | A term used to represent moral philosophy theories that inform how an individual may act in ethically complex situations. Theories include consequentialism, utilitarianism, virtue ethics, and others. |

| | |
|---|---|
| **Physiological Futility** | A concept in medicine and bioethics wherein a given intervention does not cause biological change or have any non-therapeutic or therapeutic effect. |
| **Process and Format** | A consultation methodology established by Robert Orr and Wayne Shelton wherein relevant data is assembled and evaluated prior to consultation recommendations. |
| **Resuscitation** | The attempted act of restoring one's heart function post cardiac arrest or abnormal cardiac rhythm. |
| **ROSC** | Return of Spontaneous Circulation. The goal or ultimate end of resuscitative efforts. |
| **Value Futility** | A concept in ethics wherein an ongoing or proposed therapy may or may not intend its clinical function but unequivocally fails in achieving an individuals' goals. |

# Cases

## Case 2.1

A 59-year-old man admits to his local area hospital after an aspiration event in his care home. The patient had suffered a stroke 1 month before his admission and can no longer swallow safely due to right-sided dysphagia. He receives treatment in the emergency department and transfers to the floor for continued monitoring.

Despite his risk of aspiration, the patient eats foods and drinks liquids by mouth, resulting in continued readmissions to his local hospital. The personal care home attempts to prevent the patient from oral consumption, but he still eats by mouth. The patient has a surgically placed feeding tube that provides adequate nutrition. The patient's access to foods is primarily due to his family's involvement in his care. Specifically, his family often brings food to the patient in his care home.

His attending physician places a nothing by mouth (NPO) order. Additionally, the physician informed the patient's family not to feed him by mouth and articulated the dangers of doing so. However, the patient's family and friends regularly visit the patient and bring him gifts, including homecooked foods. Despite the multiple warnings from the staff and the NPO order in place by the patient's attending physician, the patient continues to ingest small portions of soft foods orally.

The patient indicates that he is aware of the risks associated with eating by mouth but does not wish to die. The patient is willing to adhere to many safety protocols to help protect his life but still requests the care team conduct life-saving measures in the event of another aspiration event. Specifically, the patient requests a full code status in the event of cardiac arrest. This preference sparks more concern for the care team since they believe a respiratory or cardiac arrest is inevitable if the patient eats food by mouth.

The patient's family express that they do not want to harm the patient but believe he requires more nutrition to recover. They indicate that they feel comfortable feeding the patient despite the aspiration risk because they believe more nutrition will correct his dysphagia. They also wish to feed him because it is his wish to do so.

# Case 4.1

A critically ill patient with multisystem organ failure has been in the hospital ICU for the past 200 days. The patient has received various interventions, including ongoing CRRT, pressor and ventilator support, and ECMO. Additionally, she is not a candidate for transplantation. The patient is critically ill but still can understand and communicate with her care team. Although she states she is not in pain, the patient's limbs have begun to ablate, and she has developed a severe pressure wound above her sacrum.

The patient informs staff that she wants to prolong her life as long as possible and consents to any medical intervention that accomplishes this goal. Furthermore, she indicates that she wants to be resuscitated by the care team in the event of cardiac arrest. The care team is concerned about the patient's requests since they have no additional interventions to offer. Additionally, the team is worried that the currently implemented interventions are quickly becoming futile therapies.

On the 250th day of her admission, the patient's family "fire" her palliative care team. They state:

> They want to kill her! She wants to live, she said so herself! If they won't help her get better, they can leave.

Though the patient is limited in her ability to interact, she still possesses capacity. She indicates her agreement with her family with several nods. This recent development concerns the care team since they agree with palliative medicine's proposed recommendation to transition the patient to a comfort-centered care plan. The patient's dismissal of palliative care leaves the primary team without a plan that aligns with the patient's underlying goals. The attending physician determines the patient's therapies are futile and informs the patient and her family that the team will discontinue her ECMO and CRRT treatments. This determination enrages the patient's family, who subsequently take to social media to protest the hospital's decision. Furthermore, the patient's family state they intend to take legal action.

The hospital administration informs their legal and risk management departments of the situation, informing the patient's physicians that the law protects their decision to discontinue futile treatments. Still, the care team feels immense trepidation concerning the removal of life-supporting interventions.

# Contents

# Chapter 1
# A Philosophical Justification for Normative Ethics in Clinical Ethics and Medicine

Establishing components for a contemporary ethics consultation curriculum that weds philosophical analysis with analytical reasoning skills requires a retrospective analysis of clinical ethics. This task requires an analysis of the themes that clinical ethics is historically responsible for developing. One of the primary themes that readers must become familiar with includes informed consent. Despite healthcare providers' moral and legal obligation to acquire their patients' informed consent or refusal, it goes wanting for many reasons. Complacency within one's work and professional duties, provider fatigue, and lack of knowledge are all joint contributors to the problem of informed consent acquisition. Improper mechanisms of obtaining informed consent are a systemic issue and not malicious, necessarily. Instead, the lack of adequately obtaining informed consent derives from a lack of moral awareness and philosophical reasoning. We will soon uncover more themes associated with informed consent in chapter two. There, we will analyze the importance of informed consent and the philosophical justification for its acquisition by tracing a lineage of clinical ethics.

Formulating curriculum points for aspiring ethicists requires a revised iteration of ethical themes and contemporary consultation models. Educational strides surrounding consent theories and other health care ethics topics need historical landmarks that indicate the progress of moral philosophy's involvement with human health. As such, this analysis provides a summation of critical ethical concepts, methods, and texts that health care professionals should understand and implement in their practice.

## Philosophers and Health Care Providers

We first attempt to suture medicine and philosophy's divided roles by demonstrating their union's mutual benefit. Furthermore, assessing physicians' and philosophers' historical roles further aids in identifying the partnership's historical lineage. When conjoined, the physician-philosopher's role uncovers a dialectical approach to medical ethics and aids in facilitating ethical understanding across various disciplines. The merging of these roles depends on establishing a unified account of ethical practice in medicine, later defined as *common morality*. Uncovering the common morality of medical practice is significant, especially when assessing patients' needs, wishes, and goals amidst the rapid development of medical technologies. Emerging medical technologies and their influence on the complexity of medicine's ethical issues are examined in subsequent chapters. For now, we examine the interrelated roles of physicians and philosophers and work to uncover their similarities through a common morality. A primary mechanism for completing this task involves investigating the philosophical justification and basis for patient and physician autonomy. As a primary principle in biomedical ethics, autonomy justifies a standard for individual decision-making, but this may not be enough to establish a common morality. Here, we uncover the importance of the role autonomy plays in the informed consent process.

Health care possesses standards of operations regarding ethical issues and concerns. The primordial source of medical ethics derives largely from moral errors in previous research. In this respect, modern medicine has primarily learned of earlier mistakes. Safeguards like ethics committees and ethics consultations help uphold ethical concepts like informed consent and human autonomy. However, these moral themes manifest in contemporary medicine are further illuminated by demonstrating their presence in practical instances. These instances are prominent in discussions surrounding contentious topics in genetic technologies and emerging medical technologies. The genetic alteration, testing, and screening of human beings have become questionable endeavors in recent years due to their practice's moral and ethical implications and the pace at which these technologies have developed. To expand upon developing a common morality of medicine, we examine the relevant health care areas that contain some of the most pressing ethical issues.

A historical examination of the physician's role further informs the necessity and requirements for common morality in medical practice. Although commonly understood as an individual who heals, a health care provider must be examined through a multi-faceted lens to grasp their duties' full extent. Reading the physician as a philosopher informs the multi-faceted responsibilities of a physician. Many physicians' attitudes may be defined and appropriately explicated if they approach their role as healers and demand an affinity for knowledge.

Establishing a fundamental philosophy that fuses the nature of medical practice with rational analysis adds disciplined elements and criteria to the process of ethical decision-making [1]. This process aids in ethical decision-making on the physician and avoids ambiguity or superfluous practices in ethical decision-making. Regarding training health care professionals and non-professionals alike in clinical ethics consultation, the philosophical and practical knowledge curriculum points associated with the discipline require exercises in dialectical pursuits [2]. In other words, a student's understanding of the theoretical and practical concepts is more effective if they approach the philosophical aspect of medicine as another learning tool [3]. The problems associated with medical practice and ethics are vast and range from differing ideas to contentious concepts. The range of issues exemplifies the nature of the history of moral philosophy and the contemporary problems that accompany medicine today [4].

The physician-philosopher must begin with issues that coincide with medicine's praxis and subsequently implement a philosophy of experience for medical practice [5]. For this analysis, we examine informed consent and the philosophical justifications for autonomy as a primary dialectical focus for the physician-philosopher. By amalgamating the nature of medical practice as a practice of healing and the philosophical analysis of analytic problem solving, the physician may serve a professional role in achieving new medicine aims and bolster the inherent ethical aspects of their practice [6].

## A Common Morality[1] of Medicine

The physician's establishment as a philosopher is a professional amalgam of two disciplines that attempt to resolve ethical discrepancies through dialectical means. Medicine and philosophy must communicate in a common language that permits philosophy to engage in the art of medicine. This language must be consistent and understood throughout cultures, traditions, and customs. A philosophical standard is no easy task and can only develop through a shared morality. A shared character ensures that a moral norm or disposition is established within a health care ethics culture and engages with all committed persons who practice clinical ethics [7].

Due to modern medical advancements and the accompanying difficulties of new ethical conundrums, establishing a moral standard that stretches across several instances is crucial if a mutually respected ethic can develop [8, 9]. Since many medical practice variances exist, philosophy must intervene and play a central role in forming ethical principles that practitioners can abide by and advance medical practice [10]. In this respect, philosophy aids medicine as a means to ending relativistic notions and patterns of medicine. For example, who is to say what harm is or what medical treatments are excessive? What qualifies a physician to make these claims, and how do these claims impact patient rights? [11] A dialectical amalgam of principlistic judgments and practical wisdom aids in suturing the divide between philosophical principles and medical decision-making [12]. By examining ethical situations on a case-by-case basis, while simultaneously maintaining a set of principles and moral norms, ethical discrepancies may be resolved by partaking in dialectical conflict resolution [13]. This process involves defining the specifics of a case and clarifying both the language of medicine and philosophy [14]. This process aids the implementation of philosophical practice in medicine, reaffirms the individuals involved in clinical cases, and reduces conflicts between parties involved in ethical decision-making [15].

---

[1] 'Common' and 'shared' morality are interchangeable in this analysis.

The common morality serves as a universal ethical grounding that clinical ethicists and moral theorists may find helpful when dealing with multi-faceted and variant ethical dilemmas. Common morality serves as an applied moral standpoint that develops general principles, while the philosopher-physician serves as a medium through which these principles apply [16]. Although this point becomes more apparent throughout this analysis, the common morality articulated above serves as a set of principles that practitioners can only uphold through a group of individuals who understand the letter and spirit of moral objectivity. These individuals both address issues of moral authority in healthcare and aid in negotiating between stakeholders. Common morality must also become understood as a language that requires skilled interpreters and translators. In short, these curriculum points assist practitioners in becoming translators of common morality within a health care community.

## Autonomy

Promoting and respecting good medical practice requires proper acquisition of informed consent, and good acquisition of informed consent begins by acknowledging personal autonomy [17]. Achieving personal independence as a prerequisite condition for ethical practice begs the questions of how, why, and when autonomy manifests [18]. The precondition or prerequisite function of autonomy provides aid by treating persons as ends in their terms [19]. By taking charge of one's decisions and recognizing one's capacity to approach their ends as ends in themselves, autonomy becomes a reflective act that promotes an individual's ability to formulate and endorse a self-determined plan [20]. Autonomy becomes a genuine act of self-care and a precondition for ethical practice when the agent remains consistent in his plan regarding his reflections and considerations toward self and others [21]. In this respect, the act of becoming autonomous is synonymous with the health care agent's role as a philosopher. The autonomous agent amalgamates praxis and theory by taking charge of his circumstances and reflecting on their decisions [22].

Although taking charge of one's circumstances and engaging in self-reflective exercises serve as critical elements, autonomous action becomes problematic when authoritative figures threaten the ends toward which one directs themselves [23]. The physician is commonly the authoritative figure accused of hindering personal autonomy in healthcare. Theorists speculate that autonomous individuals lose their ability to remain consistent in their reflective capacities when authoritative figures, like the physician, sway the decisions of individuals who were, at one time, self-determining [24]. In this respect, the autonomous individual is never fully autonomous because external authorities' influence constantly threatens personal decision-making. Certainly, limitations to personal autonomy are not unique to only health care. As human beings, we largely adhere to moral standards constructed and developed through formal legal processes [25].

However, this discrepancy is not an issue if no significant problems exist between external sources and individuals. An autonomous individual may make independent choices even with the external influence of other individuals [26]. Following a medical authority in a situation where medical expertise is needed does not necessarily limit autonomous decision-making. Instead, the autonomous agent must objectively take the medical authority's information and decide upon a treatment or action based on proper dialectical deduction and information regulation [27].

Although various issues arise when considering the limitations of autonomous medical practice choices, the possibilities and concepts presented to an individual do not necessarily inhibit the individual's autonomy. Often, the autonomous individual encounters a problem with external influence because the agent submitting external information does not perform this task conducive to autonomous flourishing [28]. Although the autonomous individual possesses the necessary tools and skills for adjudicating various types of information, they may not do so without a conglomerate of professional individuals that aid in honing these skills. This conglomerate of skilled individuals manifests in the form of ethics committees and ethics consultants. However, ethics committees and ethics consultations serve as sects that respect individuals' autonomy so that all individuals ought to abide [29]. The need for ethics committees and consultation arises due to healthcare professionals' shortcomings, especially when obtaining consent from individuals. Ethics committees provide actions for individuals while keeping a distance to allow the autonomous agent to exercise self-care [30].

In continuing to justify the process of informed consent by establishing the role and nature of autonomy, an ontological discussion of the body is necessary for identifying the uniqueness and differentiating factors for human beings. Furthermore, an ontological analysis aids the discussion by demonstrating the role of medicine and its influence on the body. Since medicine is a practical theory of human experience, physicians are primarily concerned with remedying illnesses, healing the body, and uncovering disease causes [31]. In this respect, medicine is a praxis applied to a theory. Specifically, medicine is a practice dedicated to uncovering wisdom about human beings [32]. This act is inherently ontic by definition and requires attention due to its relevance to informed consent and autonomous decision-making.

Furthermore, the ontological facet of medical practice attempts to discover human ends and purposes [33].Medical practice's ontology is also an investigatory pursuit of human autonomy and demands philosophical attention from medical professionals [34]. Autonomous human beings pursue their ends as ends in themselves, while the nature of medicine's ontology seeks medicine's moral limitations [35].

The critical point of the ontological investigation of human beings and medicine rests upon medicine's inability to account for the value of health purely on mechanical and quantifiable terms [36].Quantifying human life in critical care requires a philosophical import to understand the complexities and difficulties accompanying autonomy and the life-determining decisions of autonomous individuals [37]. This ontological investigation of the human body and medical practice reveals that medicine, as a philosophy, unifies autonomy with personhood [38].The following section explains the importance of providing patients with ample verbal and written information regarding their course of treatment. In doing so, we uncover a practical means of upholding patient autonomy patients become fully informed before exercising their autonomy and providing clinical consent.

Historically, informed consent derives from medicine and biomedical research [39]. Issues of information disclosure, justice, and nondisclosure had been important aspects of biomedical research. However, the concept of informed consent only became prevalent in the early to mid-1970s when medical research and ethics began to focus on the physician's role as an informant for patients and research participants [40]. However, the physician's responsibility and obligation to disclose information and emphasize the quality of a patient's understanding of information opened the pathway to individuals' rights to act autonomously and choose not to partake in treatment [41].

Physicians did not universally practice informed consent at the turn of the century [42]. Physicians typically treated patients without adequately disclosing the risks and benefits of the care provided. In this respect, there was no appreciation or recognition of a patient's right to consent [43]. The physician-patient dichotomy was essentially paternalistic due to the revered expertise of medical professionals. Due to federal intervention, physicians must inform and obtain consent from patients in both practice and research. However, the term "informed" was not defined formally and thus left the practice of consent acquisition superfluously founded [44].

Finally, in the 1970s, individual liberty and autonomy became philosophically relevant topics that pertained to patient rights and medically ethical decision-making [45]. Without coincidence, the concern for autonomy, the foundation of informed consent, and bioethics' inception coincided [46].However, informed consent's history and inception are relevant to this analysis of autonomy, ethics committees, and ethics consultation [47]. The conversational paradigm between patients and physicians accompanied various difficulties in modern medicine. While the weight of conversation originally resided on behalf of the medical professional, modern clinical consent requirements in medical practice have resulted in unruly demands from patient populations. This phenomenon can be articulated further by philosophically examining the concept of informed consent. Furthermore, examining contemporary informed consent will yield a beneficial link between autonomy and authorization- a divide in honest conversation partially responsible for communication disparities in health care [48].

Realistically, full disclosure and uncovering every detail of a procedure is impossible and overwhelming for patient populations [49]. Instead, the individual must exercise their autonomy when discerning what elements of a procedure or medical intervention are necessary and relevant [50]. An uncoerced individual who understands the proposed intervention may authorize a health professional to perform a medical act upon their body [51]. In this respect, the individual who consents is exercising his right as an autonomous decision-maker by presenting a self-determining choice [52]. As this analysis demonstrated earlier, informed consent must be grounded in human autonomy. Truly informed consent cannot take place unless the preferences of patients are autonomously derived [53].

A human being's autonomous nature is undoubtedly the focal point of informed consent and the basis upon which medical practice must abide, yet relying solely on a human being's autonomy is not enough. While autonomy justifies the philosophical bases of consent, a human being must obtain the information before making an autonomous choice. The clarification of information regarding medical treatment stretches far beyond legality and into the realm of respect.

Healthcare professionals must provide information to patients in ways that are accessible and easy to understand [54]. Additionally, communicating with patients to understand their concerns and questions aids in understanding whether a patient wants to receive medical treatment or not [55]. Thankfully, various methods can aid the difficulties alongside disclosing information to patients before obtaining consent. These methods involve both oral and written information. The physician or medical liaison involved in a patient's course of treatment must relay information in a way that spends ample time explaining the details of a therapy objectively and in a manner that reduces stress or notions of obligation [56]. The medical information provided to patients intends to grant patients the necessary information to decide their care [57].

Oral information is essential to deliver to patients because a verbal depiction of a course of treatment allows the patient to hear a plan of action from a medical authority. However, the oral information is only as beneficial as the clarity of the written information. The written information should be a document that contains consistent language throughout and is easy to understand [58]. The accuracy and clarity of the written information ensure understanding for all parties involved and avoids ambiguity when patients exercise their rights as autonomous individuals by asking questions about the course of treatment [59]. The clarity of written information regarding treatment respects patients' rights and promotes their freedom to voice their concerns and opinions about a course of treatment [60].

Written information regarding a course of treatment should be accessible and comprehensive but not too overwhelming. Over-informing a patient can be overwhelming and provoke anxiety concerning the treatment [61].Treatment information should provide essential points and suggest further information, questions, topics, or concerns [62].Doing so avoids overwhelming patients with information. Synonymous with substitutive judgments used in clinical decision-making, the physician or research liaison that composes the written treatment information should include information that, in their eyes, a reasonable and responsible person would want to know [63]. In this respect, informing patients about treatments must be performed in a balanced manner that reconciles information with manageable risk and benefit analysis.

Despite the numerous standards and practices that justify medical practice ethics, the standard promoting ethical practice lies in consent acquisition. Although informed consent is an ethical means of treating patients with dignity and respect, the process of disclosure and genuinely informed consent requires maintenance of one's autonomy. By examining the instantiation and practice of upholding individual autonomy, one achieves a greater understanding of informed consent's importance and practice.

Informed consent serves as a medium that reconciles the relationship between healthcare professionals and patients. Bolstering this medium invests trust between involved stakeholders. While not exclusively viewed as a resource, trust can serve as a currency that must become delicately balanced [64]. Upholding autonomy in medical practice requires informed consent acquisition, and this acquisition must abide by specific standards. These standards help promote the proper ethical practice and provide a stronger foundation for a trustworthy relationship between physicians and patients. First, the practices of a community or sect of individuals must exercise their rights as autonomous human beings who partake in the common morality and mutually agree on the beneficial standards of ethical practice [65]. These standards, also known as the professional practice standards, may be challenging to uphold due to community disagreement or customary norms that do not abide by ethical means [66]. Second, autonomous individuals must possess the necessary decision-making capacity to evaluate information objectively [67]. This standard is pertinent to individual autonomy because respect for an individual's decisional ability upholds autonomy's self-care aspect [68]. Patients must also determine if they have received an appropriate amount of information regarding a course of treatment [69]. Since patients require different forms of care, adequate information regarding treatment may fluctuate and may not be quantifiable [70].

Although the standards for informed consent listed above are practical, they still contain various issues. Informed consent is essential for aiding ethical practice in medicine. Furthermore, informed individuals can make decisions for themselves, direct their lives according to their will, and not allow the influence of others to persuade them to partake in practices contrary to their morals [71]. However, three vital components of informed consent require examination. This examination acquisition of consent and the role autonomy plays in this process. These components include informing the individual of the details of a course of treatment, informing them that they have the right to voluntarily choose to accept or decline treatment, and performing a comprehensive assessment of their capacity [72].

Promoting a patient's autonomy requires rational deduction of their circumstances and the ability to make voluntary deductions. However, excessive information about a course of treatment may be detrimental and burdensome to one's self-care. For this reason, it is essential to establish a disclosure method in which health care providers can deliver a detailed yet accessible explanation of a course of treatment [73].Naturally, these instances encounter various complications and require resolution from individuals trained in value facilitation. Medical professionals must thereby rehearse their abilities to relay complex information in a manner that accommodates a given patient or surrogate in their situation. In this respect, the proper presentation of information upholds autonomy and solidifies the necessity of ethics consultation [74].

## Coercion and the Vulnerable

Medical professionals and ethical authorities must make efforts to avoid the exploitation of patients [75]. Exploitation and other autonomy threats violate the common morality by not abiding by the body's ontological justifications discussed above. Synonymous to Kant's moral theory and the *telos* principles of autonomy, treating human beings as a means to an end rather than ends in themselves do not adhere to bioethics' canonical principles [76]. Exploitation in medicine may professionals meet specific criteria. First, patients are in jeopardy of succumbing to unethical practices if the treatment proposed mutually benefits the healthcare professional [77]. Second, exploitation occurs if the mutual benefits violate the integrity or authenticity of consent acquisition [78]. These two criteria demonstrate patients' exploitation by treating them as means to ends rather than ends in themselves.

Obtaining consent from patients can undoubtedly serve as a medium for exploitation via improper acquisition. This analysis discussed how informing participants correctly allows individuals to exercise autonomy by making informed decisions about their treatment course [79]. However, despite the autonomy human beings possess, various individuals are especially susceptible to exploitation. These individuals include the economically disadvantaged, the disabled, and the elderly [80]. These individuals are especially susceptible to exploitation because of their economic, physical, and socially discriminated situations.

Additionally, these individuals may lack access to comprehensive health care and may have never received adequate treatment [81]. These individuals are susceptible to exploitation in research and clinical care because they may not have access to proper information before giving consent [82]. For instance, the written information that ought to be provided to the patient before treatment begins is an essential part of informed consent. However, this information is useless if the patient is illiterate or unable to comprehend the information orally. These individuals are autonomous but cannot exercise their right to autonomy in healthcare due to inhibiting factors. Simply excluding these groups from medical treatment is not an option because it would be unjust and discriminatory.

Excluding or coercing these individuals into a treatment plan may not understand or agree to violates their autonomy and the physician's role as a healer and philosopher [83]. Aiding these groups by ensuring proper understanding before treatment commences is of the utmost importance. While competent patients in vulnerable groups can learn their course of treatment, patients who lack capacity and a proxy must have access to a group of individuals who will make decisions in their best interest. The role of an ethics committee or ethics consultation is significant regarding the prevention of exploitation of individuals in vulnerable situations [84].

Coercion presents itself as a unique violation of autonomy and a detriment to the process of informed consent. If a physician coerces a patient into consenting to treatment, then the consent is not valid [85]. Barring occasional exceptions regarding paternalistic determinations by physicians as a necessary action, coercion more often manifests as a detrimental element to autonomy and a violation of informed consent [86]. Coercion, along with other issues mentioned in this discussion, is an issue that requires the implementation of normative methods in healthcare ethics through the medium of ethics committees and ethics consultation [87].

Coercion is unique because it involves the dangers of sufficient harm that may befall a patient and thus strips away their freedoms and right to choose [88]. The pressures a physician or healthcare professional may put on a patient must be limited to the benefits and harms involved in the course of treatment and must not include threats to discontinue treatment [89]. The above section on proper written consent is of the utmost importance when addressing the issue of coercion [90]. If consent solely relies on the oral delivery of information, coercion is undoubtedly more likely to occur [91].

## Paternalism

Paternalism in healthcare stands out as a more pressing issue requiring ethical solutions due to its pervasive and authoritative nature [92, 93]. Although paternalism is no longer a default attitude in healthcare, confusion exists surrounding its status and when professionals are morally justified in its' practice. Furthermore, many healthcare professionals are unaware of what instances qualify as instances of paternalism. This section examines the harms and benefits of paternalism and the role ethics committees can play to regulate this practice [94].

Paternalism intends to benefit another human being, yet its practice still requires moral justification. Value judgments, processes, or proceduresimpose upon a human being who may not necessarily request or agree to a proposal [95]. Paternalism is an example of a practice that shows that good intentions or motives are not sufficient conditions for acting morally [96]. This point asserts that good intentions are not the only element needed for moral decision-making.

Moreover, additional elements require assessment before a moral judgment or action may take place [97]. Defining paternalism is complicated when considering the ambiguity of discerning morally acceptable actions. Identifying instances of warranted paternalism requires the amalgam of the physician and philosopher. Providing a comprehensive definition of paternalism also justifies the autonomous nature of informed consent and the need for ethics consultation because regulatory standards of justification require adjudication by a team of individuals who understand normative ethical principles [98].

Determining a paternalistic act requires identifying the act as one that must direct itself toward a positive end for the affected individual [99].A comprehensive definition of paternalism must include language that recognizes paternalism's positive and negative implications. In this respect, identifying instances of paternalism requires four criteria: (i) The paternalistic agent believes their action benefits the recipient, (ii) the paternalistic agent recognizes that their actions require moral justification, (iii) the paternalistic agent believes obtaining consent from the recipient is not possible or assumed, and (iv) the paternalistic agent believes the recipient has the utmost confidence in the paternalistic agent's decision [100]. The first criterion justifies the paternalistic act if the agent intends to benefit the patient directly [101]. The second criterion is synonymous with parents making decisions for their children. A paternalistic action is partly justified if the agent's actions benefit individuals other than themselves [102].

Although paternalism arguably violates various moral standards, it still must become normatively justified. It is not because paternalism is a directly immoral approach to decision-making, for this notion is false. Instead, paternalism is yet another tool available to medical decision-making practices if the situation meets necessary conditions. Practitioners of paternalistic decision-making should implement this methodology of decision-making as a conditional decision-making tool.

## Emergencies

Emergencies inherently require a paternalistic model of decision-making. Whether persons are obtunded or otherwise, most emergent situations warrant providing treatment over objection. For instance, if a person needs CPR and a bystander knows how to perform CPR, the bystander is ethically justified in performing CPR since the emergent situation implies the patient's consent.[2] The bystander acts ethically supportably because they believe their action is a reasonable means of aiding someone requiring medical attention [103]. The last criterion implies a level of dominance or dominion over the recipient because the paternalistic agent believes the recipient can still make their own decision on a specific matter. However, the paternalistic agent believes their decision is a beneficial choice for the recipient [104].

One of the more significant concerns surrounding paternalism is the necessary usurpation of autonomy for its implementation. The nature of paternalistic decisions diminishes one's autonomy while bolstering another agent. A medium between these two elements must exist for paternalism and autonomy to coexist in a beneficial union. Patients indeed have the right to receive information about their treatment plan and make judgments or decisions based on the information they receive [105].Autonomy remains the focal principlesurrounding individual liberty. Unfettered adherence to patient autonomy hindershealthcare professionals' authority and provokes paternalistic solutions [106]. Regardless of how paternalism manifests, paternalism always includes a hindering of one's autonomous choices [107]. Paternalism becomes a further problem in medical practice when examining principles of nonmaleficence and beneficence [108].

---

[2]Emergent conditions in this context refer to those individuals who experience a trauma or need immediate medical attention for a chronic condition. Emergent situations in this context do not refer to end-of-life cases.

Historically, nonmaleficence and beneficence serve as foundations for paternalistic decisions. However, imposing procedures and treatments upon individuals who can still make decisions for themselves render their autonomy valueless [109]. Still, many theorists argue that there exist various instances where paternalism is warranted, and thus undermining human autonomy is justified. For example, a care team imposes a life-saving treatment on a patient who had refused [110]. In this instance, the care team usurps the patient's autonomy. Still, it may be reasonable to deem the patient's autonomy as afflicted or diminished in live threatening instances. Since the patient is incapable of exercising their autonomy to promote their flourishing as a human being, paternalistic means of treating the patient may be warranted.

Another example where personal autonomy becomes hindered by a paternalistic approach pertains to the decisions of parents who are Jehovah's Witnesses. If the infant child of a Jehovah's Witness requires a life-saving blood transfusion and the parents refuse based on their faith's tenants, it is morally justifiable not to adhere to the parent's wishes. Though the patient's parents have good intentions and have prioritized their faith, their request asks medical professionals not to treat a vulnerable human being. This instance is an example of paternalistic decision-making over a surrogate's wish. The life-threatening nature of the situation justifies a care provider's treatment over surrogate objection, primarily since the justification for withholding life-sustaining interventions is rooted in religious belief. A minor cannot consent because they have not developed cognitively enough or experienced enough phenomena to make self-determinations regarding a treatment decision. However, the undeveloped or incapable of minors and incapacitated persons alike justifies treatment over objection, particularly in emergency settings. The parents raise their child under Jehovah's Witness practices. Still, they have not autonomously made their religious determinations at this stage of their life [111]. While physicians and other caregivers may assert paternalistic notions upon patients, it is the job of ethics committees and consultants to identify and analyze situations objectively, facilitate parties involved, and aid individuals on morally acceptable treatment and care options [112].

Thus far, we have uncovered physicians', healthcare professionals,' and ethicists' shared roles in becoming moral agents in clinical settings. We have also revealed the ontological justification for autonomy and the importance of informed consent as one of healthcare providers' ethical obligations. The specific problems that result from morally questionable cases involving autonomy and informed consent include exploitation, coercion, and paternalism. It is vital to understand how normative methods in ethics manifest and remain enforced in contemporary clinical situations. The following section discusses ethics committees' manifestation and ethics consultation and their pertinence in current medical practice. Furthermore, analyzing various issues surrounding ethics consultation justifies their approach. This analysis demonstrates that these practices are rooted in a philosophical basis of medical practice and uphold moral excellence.

# Historical Precedence

The results of the Karen Ann Quinlan case of 1976, the New Jersey Supreme Court concluded that ethics consultation must become a necessary component in ethically or morally questionable situations in hospital settings [113]. This event prompted these committees and individuals within these committees to partake in decision-making processes. These decisions require collaboration alongside physicians, family members, patients, and hospital administration [114]. Due to the Quinlan case, ethics committees' initial role involved substitutive judgments with incompetent or incapacitated patients. Quinlan's case was monumental because ethics committees grew beyond mere advisory panels, commonly cast aside during decision-making processes [115]. Following the Quinlan case, the role and legitimacy of ethics committees were slow to develop. However, the novelty behind the integral decisional involvement of ethics committees leads directly to developing commissioning bodies like the Joint Commission on Accreditation of Healthcare Organizations (JCAHO) [116].

Although the Quinlan case granted tremendous priority to ethics committees and partially relieved physicians of the burden of decision-making, the instantiation of ethics committees aided patients by giving them a panel of advocates that would advise and assist these individuals during treatment [117]. Understandably, the integral priority ethics committees obtained sparked controversy among the medical and philosophical community. One of the primary issues surrounding contemporary clinical ethics committees involves applying medical professionals to the ethics team [118].

While various theorists believe only philosophers should be members of the ethics team, other individuals believe medical professionals should be involved. Conversely, many individuals suggest philosophers have no business in medicine [119]. For example, a profession like neurosurgery is a high-risk specialty that involves various issues that may accompany specific procedures and treatments that leave patients in physical and mental states that are emotionally taxing for families [120]. With this point in mind, should neurosurgeon-bioethicists partake in an ethics consultation? This question is especially relevant when addressing the issue of amalgamating the roles of medical professionals and philosophers. This analysis argues that the amalgamation of these roles aids the moral fortitude of maintaining personal autonomy, respect, and patient advocacy in and through ethics consultation and ethics committees. However, a physician's role in ethics consultation is a highly beneficial asset to decision-making processes by providing ample information about specific cases involving knowledge out of the philosopher's scope [121].Medical professionals have advanced knowledge of a medical issue. The philosopher has ample knowledge of ethics and normative theories. Case-by-case evaluation with both professions involved is a relationship that could combat the complexity of medical situations and aid in viewing medical issues more clearly [122].

This analysis has primarily examined informed consent issues and the ethical demand for a philosophical basis for medical decision-making. The above analysis describes the necessity of ethics committees and consultation due to the need for a medium for upholding philosophical assessments in medical practice. Furthermore, physicians' and philosophers' union serves to justify and bolster ethical practice effectiveness in medicine. Although there exist various issues surrounding informed consent and the philosophical implications of its acquisition, medical professionals are making significant efforts to suturing the divide between medicine and philosophy by upholding new standards of care and practice. Specifically, the future of ethics consultation is promising if it accepts the amalgam of disciplines and works to implement case-by-case practices [123].

One of the primary issues surrounding ethics committees involves a lack of expertise from the various members that comprise the committee. The key components this text presents can aid this knowledge discrepancy by implementing the expertise of professionally trained ethicists. While an ethics committee's responsibilities are essential, it should not be the committee's responsibility to conduct ethics consultations, nor should it be the committee's responsibility to hold family meetings when addressing contentious conversations [124]. Tasks involving alleviating moral quandaries, facilitating dialogue, and using philosophically relevant literature to justify ethical recommendations should come from professional ethicists. However, these assigned responsibilities should not discredit the role of an ethics committee. Modern ethics committees can perform efficiently if they successfully serve two functions. First, ethics committees should work to disseminate catered ethics education throughout a health care system. Second, ethics committees should act as venues for quality improvement projects.

Education development and quality improvement projects identify an issue that the committee deems a necessary point of correction within the health system. The committee subsequently takes ownership of the project and tracks quality improvement through a longitudinal study. While these tasks are uncommonly associated with ethics committees, the redistribution of roles and responsibilities within ethics committee structures bolster the effectiveness of disseminating ethics throughout a health care system and aid in utilizing professional resources in ethics [125]. Although methods for supporting the practice and acquiring consent have been reevaluated and assessed, issues are still present when attempting to obtain consent in a morally stringent manner. No unifying rule exists that resolves ethical issues in medical practice [126]. However, rather than developing a single rule, amalgamating medical practice and philosophy into an instantiation of rules and formulations that aid in ethical theory is a more tangible and accessible viewpoint to ethicists and healthcare professionals [127].

## Situational Ethics and Moral Obligation

Adhering to maxims lies at the heart of Kant's insights on moral obligation [128]. However, deontological maxims are rigid and have a propensity for stagnancy when dealing with multi-faceted ethical situations. Since ethical dilemmas manifest in many different forms and entail various factors that contain several variables, each moral problem requires individual attention. A possible remedy for the unpredictability and lack of uniformity in ethical situations involves an approach known as situational ethics. Situational ethics is an honest approach that recognizes the vast differences in moral problems and addresses them accordingly [129]. This approach is beneficial if the philosophical basis for informed consent is rooted in autonomy. Situational ethics is unique because it does not limit its practice to a singularity and does not apply standards and ethical norms to all situations [130]. Since this method relies on a philosophical basis for medical decision-making, situational ethics works out of affection and empathy for individual plights [131].The relationship between philosophers and physicians and their coexisting role in ethics committees makes normative theories and bioethics methods applicable in contemporary situations when situational ethics demands a proper union of these disciplines [132]. Situational ethics is just one of many novel approaches to ethical decision-making in healthcare. If professionals give individual attention to each ethics case rather than merely apply and reinforce ethical norms, additional care and concern for human autonomy occurs. Situational ethics can further aid future medical pursuits by protecting patients' rights by developing a more comprehensive method for informed consent, anonymity, and confidentiality [133].

Including trained medical professionals who possess knowledge of their specialty and bioethics provides helpful information that contributes to ethical discussions when resolving moral disagreements, family disagreements, treatment plans, and hospital protocols [134]. Furthermore, physicians trained in bioethics, or succeed in amalgamating medicine and philosophy, bolster the consultation process in their specific specialty [135].

Informed consent and autonomy issues can resolve in productive and manageable ways by forming ethics committees with individuals who exemplify knowledge in their specialized field of medicine and are informed about the various bioethical issues in healthcare. This portion of the analysis shows the critical role philosophical analysis plays in ethical decision-making in medicine, and ethics committees and consultations uphold ethical norms and theories. However, applying the theoretical and historical framework outlined in this discussion is only relevant if these themes apply to ethically appropriate areas of medical practice. The analysis bolsters the implementation of bioethical theories by addressing biomedical technologies and modern technology in contemporary health care. The research demonstrates the effectiveness of ethical interventions and their relevance among various curriculum components for ethicists.

## Genetics and Ethics

The discussion turns to the ethics of medical genetics, clinical genetics, and the technologies entailed in both fields to address autonomy, informed consent, and paternalism. Giving these contemporary practices attention provides a medium to which the ethical themes discussed in this chapter may apply. Additionally, the discussion distinguishes between medical and clinical genetics. Addressing contemporary medical and clinical genetics practices elucidates modern medical procedures' appeal and benefits and clarifies the boundaries that modern technology may or may not ethically permeate. Addressing ethics and genetics, in turn, yields a preliminary discussion of ethical permissibility, which provides a foundation and demand for healthcare ethics consultation.

Furthermore, this discussion further supports the need for additional curriculum components for clinical ethics consultants. Despite the literally billions of variances of genome patterns in humans, the similarities of human genomic structures are remarkably close. In 1990, geneticists and scientists began mapping, sequencing, and storinga complete human genome's genetic information [136]. In 2003 the Human Genome Project (HGP) succeeded in accomplishing its task by mapping approximately three billion nucleotides and identifying all protein-coding genes within these nucleotides [137]. This accomplishment opened various doors in the diagnostic world by providing possibilities for specializing diagnostic treatments for patients and prenatal health [138].

## Ontology, Transhumanism, and Human Genetic Screening

Human genome mapping is an accomplishment that redefined and bolstered clinical genetics from diagnostic options to treatments and prevention. Although the reserve of tools unveiled by the HGP holds relevance for both medical and clinical genetics, it is crucial to make a professional distinction between these two fields. While medical genetics focuses on the history and prevention of genetically-based diseases, clinical genetics pertains to medical genetics's application to diagnostic procedures, prognoses, and hands-on treatments of genetically inherited diseases [139]. With the discovery of these valuable tools, clinical geneticists can now use personalized maps of a patient's genome to aid in diagnostic ventures [140]. The HGP provides exciting insights into the composition and pedigree of human beings and aids in customizing human beings' therapies, expediting effective treatments. With the information gathered from the HGP, clinicians can now target patients' unique biology and physiology to enhance patient care [141].

Personalizing patient care with genome mapping aid may prevent or delay diseases, reduce mortality, reduce disease incidence, and prescribe proper treatments and medications with less trial-and-error [142].Genetic testing has allowed researchers and clinical geneticists to cure and prevent genetically based ailments that were never before possible in recent decades. However, the same information gathered from the HGP may also be used for effective genetic testing and screening methods. Genetic testing and screening is perhaps the most widely used medium for clinical geneticists [143]. The analysis examines these tests to elucidate the uses of genetic technologies.

Furthermore, examining the variances in genetic testing demonstrates the effectiveness of detecting diagnostically relevant human genome signs [144].One of the tremendous medical benefits derived from the knowledge attained by completing the HGP in 2001 is the development of testing and screening for genetic abnormalities within the human genome [145]. Advanced diagnostics has migrated from the realm of science fiction to reality at an alarming pace. In 1956, geneticists established the modal chromosome number of 46. In 1959, experts identified the trisomy-21 imbalance, thus uncovering an explanation for Down Syndrome, and, in the mid-1980s, molecular genetic techniques developed into a field of professional diagnostics [146].

The excitement surrounding the fascinating developments in clinical genetics often overshadows the alarming pace at these discoveries manifest. Genetic technologies are no different from other developments in modern technology in that the findings and innovations surrounding the field are uncovered and implemented swiftly and often without philosophical review [147].Impactful medical advancements in genetic technologies develop quickly, but the speed at which development usually provokes various ethical dilemmas and situations requiring recourse. However, despite the pace at which these technologies are discovered and used, many diagnostic genetics discoveries have proven extraordinarily beneficial [148]. Genetic testing is the diagnostic evaluation and analysis of DNA [149]. Since DNA possesses a tremendous amount of information, genetic testing ranges from chromosomal analysis to gene linkage, *in situ* hybridization, to gene sequencing [150].

The variances in genetic testing share a common goal in clinical genetics, namely, the purpose of identifying a specific genetic cause, etiology, medical malady, or condition. These tests differ based on the genetic variance of a particular abnormality [151]. For instance, carrier testing involves the genetic testing of an individual who is not affected by a genetic condition but still possesses the genetic abnormality in their DNA. This abnormality could potentially transfer to the individual's offspring. In contrast, prenatal testing pertains to the identification of congenital abnormalities and changes in an unborn fetus [152].

The tests mentioned above yield tremendous benefits for patients and would have been impossible if not for the discoveries uncovered in the HGP. The discussion previously mentioned the profound rate of technological development in health care. Although the benefits of genetic testing are apparent, the pace at which these tests develop yields risks of ethical questionability. For instance, prenatal testing requires access to cells removed from the developing embryo, while cells from amniotic fluid require acquisition during amniocentesis [153]. The practical issues involved in genetic testing are of the utmost importance concerning their use in clinical medicine. Issues like cost, ethical considerations, and test completion details are all integral parts of genetic permissibility [154].

Furthermore, genetic testing may not yield the desired answer for clinicians and patients. This issue becomes especially problematic when certain boundaries are crossed to perform a genetic test in the first place [155]. In terms of genetic screening, or the process of genetically searching for predisposed ailments in patients, tremendous ethical issues arise in patient discrimination. While one group of individuals may not be prone to a genetic disease, another group may be unjustly discriminated against for abnormalities in their genetic makeup [156].

Genetic testing and screening issues become overlooked due to the rate at which genetic technologies develop [157]. Simply because human beings can act does not necessarily mean they should perform that action. It is also important to note that modern technology's rapid expansion is not isolated to genetic technologies alone. Technology has developed so quickly in recent decades that it has replaced itself in nearly every facet of its existence. The following section implements Martin Heidegger's philosophy in his essay The Question Concerning Technology to discuss other genetic technologies(*Die Frage Nach der Technik*) [158]. To uncover man's relationship with technology, Heidegger reveals the dangers associated with technology. For this investigation into genetic testing, screening, and the necessity of ethics consultation, Heidegger's philosophy will aid as a medium that uncovers the causes and dangers of rapidly expanding genetic technologies. Genetic technology's tremendous strides in genetic testing and screening seemingly yield fruitful human beings' benefits. Performing genetic screening for congenital abnormalities and subsequently providing helpful treatments aids in customizing personalized diagnostic treatments, while genetic screening bolsters the effectiveness of preventative medicine [159]. Although genetic testing and screening share expected benefits, the ethical implications of their practice become further illuminated by Martin Heidegger's philosophy in his essay *The Question Concerning Technology*.

Heidegger's philosophy is used as a medium to uncover the underlying ethical issues of genetic testing and screening. Furthermore, Heidegger's essay aids the discussion by emphasizing the dangers of genetic modification and the future of genetic technologies. Heidegger's controversy primarily demonstrates the implementation of philosophical aids for contemporary curriculum components and shows the benefits of combining health care professionals' and philosophers' roles. The following section details the ethical impact of modern technology on nature by examining the relationship between nature and modern technology, the imposition of modern technology has on nature, and the danger modern technology places on nature [160].

Medical professionals rarely scathe modern technology or negatively discuss its development when referring to modern medical technologies. Devices like the MRI that revolutionized diagnostic medicine are tremendous benefits to human health. Current technology's power shifts from its practical use to a tool of abuse quickly when technologies implement into actions that manipulate human makeup [161]. Since genetics involves the natural molecular composition of human beings, modifying and tampering with genetic structures raises various ethical issues. Although Heidegger does not refer to medical technologies in his philosophy, he does give a detailed analysis of man's abuse of nature when developing and using modern technologies. If addressed appropriately, Heidegger's philosophy and the caution he provides within his technological assessment yield fruitful insights into the dangers of tampering with human genetics in contemporary practices [162].

Heidegger describes modern technology's relationship with nature as dominating, imposing, exploitative, and forced [163]. However, the term that Heidegger stresses the most during his modern technology assessment is challenging [164]. It is this term that separates two kinds of technology, handicraft technology, and modern technology. While handicraft technology is a form of development and enhancement that involves man's cooperative conjunction with nature, modern technology demands satisfaction and forces and challenges nature to yield to man's demands [165]. In this respect, nature works *for* instead of *with* humankind [166].Treating humanity as a means to an end is especially problematic when referring to man's relationship with nature since man is inherently part of nature [167].

Once the dichotomy between man and nature breaks by demand for nature's submissiveness, man's relationship with nature transforms into[3] a standing reserve forced to provide for human needs, whims, and satisfaction [168]. Heidegger considers man's relationship with modern technology a dishonorable challenge since nature has no means of protecting or defending itself from modern technology's violent demands [169].

Modern technology's imposition upon nature is of utmost importance when examining contemporary technology's ethical permissibility in genetic alteration. Heidegger describes modern technology's imposition on nature as an unwarranted advantage [170]. When man attempts to reap unnatural bounties from nature deliberately and forcefully, man begins to impose himself upon what is natural. In the Heideggerian sense of the term set, man's relationship with nature is synonymous with playing God [171]. Heideggerian imposition is especially relevant when discussing the ethical permissibility of genetic alteration. While nature determines the genetic makeup of a human being, the modification and tampering with genetic makeup are examples of man's unwarranted and forceful demands from nature. Rather than working with humans' naturally occurring biology, genetic alteration and modification exemplify the unnecessary need and imposition modern technology places upon nature [172]. For instance, it is not uncommon to pray before consuming the bounty when traditionally reaping farmed crops from the land. However, if the same crops are genetically altered and modified, a prayer to God before their consumption is almost unwarranted due to man's imposition upon nature [173].

Heidegger notes that the primary danger inherent in modern technology lies in the possibility of losing human freedom [174]. Since the abuse of nature stems from the dependence and imposition of modern technology, human enslavement is a genuine opportunity to reliance on modern technological advancements. Suppose man's relationship with modern technology is one that continues to sap nature of its resources by forcefully modifying and disrupting its homeostasis. In that case, human beings may relinquish their former free-standing relationship with nature by becoming inherently dependent on the newly fostered artificial means of growth [175].Consider the following example: If human genetic alteration becomes common and most human beings become genetically altered, nature no longer coexists with man and instead works *for* man. In this respect, man becomes enslaved to his

---

[3] Enframed, as enterpreted from the german conception of '*Gestell*'.

creation since humankind cannot thrive without the newly modified unnatural human biology state [176].

The discussion of Heidegger's philosophy and its relationship with modern genetic technology intends to caution against natural manipulation. While genetic modification and alteration may categorize as modern technology, in the sense that its practice inherently abuses nature, genetic screening and testing may manifest as handicraft technology or a technology that works with nature. However, genetic alteration benefits, *i.e.,* therapeutic modifications to prevent inherited diseases or abnormalities, are still categorized as modern technology. Before investigating the permissibility of genetic testing and screening, a brief consideration of the philosophy of transhumanism and transhumanism's effects on modification technologies requires examination. There exist schools of thought that both advocate for and caution against the modification of human beings. Nevertheless, the philosophy of modification is a school of thought that aims to bolster the varying perspectives entailed within the realm of human modification. While the philosophy of Martin Heidegger serves as a medium that explicitly cautions against the manipulation of nature or, for this investigation, the manipulation of human genomics, the philosophy of transhumanism attempts to seek out the enhancement of human intelligence, physical performance, and natural resistance their current limitations [177].

It is essential to note misconceptions that transhumanist philosophers attempt to rebut. Philosophy tends to explain its practice's nature by explaining what it is not rather than what it is, and transhumanism is no different in this respect. First, transhumanist philosophers argue that it is not the goal of a transhumanist philosophy to achieve a perfect human subject. Instead, transhumanism seeks to perpetually improve what they consider to be a flawed piece of biological engineering. Contemporary phenomenologist and philosopher of technology Don Ihde counters this argument by stating that transhumanism seeks to develop a futuristic utopia of human beings. However, transhumanist literature combats this criticism by explaining that the actions entailed in transhumanism, synonymous with the Aristotelian conception and use of *Energia* (ἐνέργεια),[4] seek to perpetually improve and engage in an act that inherently entails an end within itself [178]. Although it is impossible to achieve human perfection, the very act of attempting human perfection by expediting the natural model and construction of human beings is first, a bold and arrogant claim against the natural, and second, a dangerous step in segregating different types of modified and non-modified human beings [179].

A second misconception involves transhumanist philosophers' notion of despising or loathing their bodies and human beings' current biological makeup. Furthermore, this criticism serves as a primary reason why transhumanist philosophy seeks to expand the current state of human composition [180]. This point is especially relevant to Heidegger's risks in his philosophy in that the disruption of, rather than cooperation with, that which is natural yields dangerous and unethical results. Although transhumanism advocates may not loathe their bodies, they believe

---

[4] *Energia* (ἐνέργεια), from the Aristotelian conception of 'actuality' or 'ultima.'

improvements are critical to an otherwise flawed piece of engineering by developing a human body resistant to aging, disease and capable of inhabiting different bodies, *i.e.,* virtual bodies [181].

The kind of modification entailed in transhumanism fits Heidegger's technology philosophy, namely, an inherently modern technology. The contemporary technologies discussed by Heidegger and transhumanists alike involve manipulating and disrupting what is naturally occurring in biological structures. Human beings' biology is affected by implementing drugs, procedures, and treatments, human genomics' direct modification may classify as an unwarranted and unethical manipulation of natural creation. While modern medicine pertains to handicraft technology, in that skill is amalgamated with an art form to produce beneficial results, the modification of human genomics for the "betterment" of humanity directly manipulates the natural and thus deemed, in the Heideggerian sense, as modern technology [182].

Despite the pervasiveness of human enhancement and the editing of human genomics for biological improvement, it is not unreasonable to examine human genomics' modification as a beneficial tool for human therapy. While this form of editing human genomics is still ethically questionable, it is necessary to discuss genetic ethics consultation. While modern technology reflects a deliberate and pervasive kind of manipulation to that which is deemed natural, handicraft technology is a medium that attempts to uphold a legitimate relationship with that which is considered natural [183]. In terms of genomics, handicraft technology includes the various methods and technologies associated with therapeutic genetic testing, screening, and prevention. Investigating the intricacies of ethics consultation surrounding therapeutic genetic testing and screening requires a brief consideration of therapeutic screening and testing methods and procedures [184].

Arguments and policies surrounding the promotion and use of genetic screening for newborns may be traced back to a case in 2000 involving Ben Haygood. He died due to a rare metabolic disorder known as medium-chain acyl-coenzyme (MCADD). MCADD requires those inflicted to avoid long-term fasting due to increased risk of disease and death [185]. Advocates for genetic screening in newborns argue that a simple genetic test that Haygood did not receive could have prevented his death. Furthermore, the same advocates argue that various alternatives for improving children's health are rooted in genetic screening, and these tests must become systematized in healthcare as policy standards [186]. However, concerning the philosophical justification for technological use in healthcare discussed above, its relationship with nature must determine genetic screening's ethical implications. Although there is a push for the policy-mandated screening of newborns for all possible congenital abnormalities, only some ailments require mandated testing [187]. Despite promoting comprehensive genetic screening in newborns, some screening methods for illnesses and treatments may not coincide with handicraft technology's ethical permissibility and work against the naturally occurring human body structures. In Ben Haygood's case, the screening method for MCADD involves a simple blood test that does not manipulate or disrupt his genome's structural integrity.

The treatment for MCADD simply involves eating a scheduled diet to ensure the body's proper production of glucose, which also does not disrupt or interfere with the human body's genomics [188]. In Haygood's case, both the screening method and resolution to his ailment would qualify as quintessential examples of diagnostic medical interventions, despite human genomics' involvement. However, state law mandates only specific genetic screening methods, including PKU, sickle cell anemia, congenital hypothyroidism, and glucosemia [189].

The primary screening method for newborns involves a heel-stick blood sample drawn 24–48 hours after birth. However, tandem mass spectrometry, a technology discovered in the 1990s, has become a far more efficient and effective way of screening newborns for PKU, cystic fibrosis, and even MCADD. Despite its effectiveness, this technology must undergo ethical scrutiny and supervision regarding the effects these interventions may have on human beings [190]. Tandem mass spectrometry measures the levels of metabolites in human blood. From this measurement, scientists may determine what kind of metabolic disorders are present [191]. Since tandem mass spectrometry uncovers a result and presents physiological options for treatment, one can argue that this form of screening is ethically permissible [192].

While the human body cannot detect ailments that may dwell within, the addition of a microchip that detects human genetic disorders qualifies as a human modification. In this respect, the delineation between modern technology and handicraft technology in healthcare becomes clear. The effectiveness of chip technology does not yield more fruitful results than current tests except for early detection instances. While the heel-stick method and tandem mass spectrometry detect genetic ailments via medical tests, augmentation through chip technology skews both the intent and agency of medical procedure. Additionally, this kind of modification tends to yield unfair advantages to individuals who have access to this kind of human modification [193].

The section above demonstrates the philosophical differences between screening technologies and augmentation and preventative screening. Naturally, these aspects of genetic screening and testing possibilities are fundamental considerations when receiving genetic counseling. However, these considerations often go wanting due to the contemporary demand for practicality. Philosophical considerations regarding ontology and the permissibility of procedures and technology are definitive points of argumentation in contemporary health care ethics. Clinical ethics consultation must curtail individuals' concerns and situations. Some final practical considerations require attention before delving into the various theoretical approaches to genetic counseling that aid decision-making and ethical facilitation.

Genetic testing and screening are incredibly unique consultation topics due to the inherent necessity of performing these procedures, typically during the early stages of life. However, many individuals are faced with onerous predicaments due to this necessity, including cost, fair use, confidentiality, and consent. An examination of these issues provides insight into the conversational and counseling components of an effective genetics consultation.

Although cost is a primary issue concerning genetic testing, various public screening policies alleviate financial burdens. Furthermore, various state legislations require mandatory minimum genetic screening for all newborns.

State-mandated required screenings involve screenings for PKU and sickle cell disease but do not necessarily cover genetic diseases [194]. In Bed Haygood, MCADD was not a disease requiringmandatory screening and thus cost more to perform this procedure voluntarily [195]. Although the design is simple, screening for genetic diseases other than state-mandated screenings is costly.

Furthermore, due to the increased cost of non-mandated state screenings, parents often do not opt for voluntary screenings. Cost considerations for genetic screening are not a minor concern of parents. Most genetic diseases, unless otherwise specified in family history, present no motivation for preemptive screening. Although families have the right to choose what screening procedures their newborns receive, many life-threatening or life-altering diseases go undetected and untreated. Rather than risk usurping a family's autonomous decisions for their kin by imposing more state-mandated screenings, a well-comprised, comprehensive ethics consultation that presents all relevant genetic screening options to a family is a far more formidable option. In doing so, viable alternatives that coincide with a hospital's mission and goals receive adequate facilitation [196].

The massive amount of information genetic screening and testing provides is not limited to the individual undergoing the screen or test. Since genetic science specifically pertains to genomic lines of pedigree and familial traits, multiple family members in the same genomic sequence can receive the same genetic diagnosis. For this reason, understanding one's family history is extremely important. In 2004, the U.S. Surgeon General's Family History Initiative urged families to become familiar with diseases prone to inflicting their family line. However, family history is a complex facet of information to regulate on an individual basis. One family member's provision of their family's history is also shared with all the same genomic line members. In this respect, issues of privacy and fair use in interpreting genetic information become problematic. While one family member may want to provide their history with their families' history for diagnostic purposes, the same family member inherently discloses information that may not be theirs to distribute [197].

A patient who receives a genetic diagnosis may also learn that their disease is present in another family member that did not request a genetic screen. Breaching confidentiality for the sake of preventing harm becomes a severe ethical concern when performing genetic screening and testing. This point is especially relevant for ethics consultations regarding genetic screening. Since genetic screening affects more than just the one patient undergoing the screen, ethics consultants must serve their purpose as facilitators to bridge communication gaps between previously uninvolved party members. Since no U.S. authority exists that mandates information disclosure in genetic incidents, ethics consultants must provide options and encouragement for patients afflicted with this onerous task [198].

A final practical consideration regarding genetic ethics consultation pertains to decision-making capacity. Many health care procedures may require consent from an individual who cannot do so, and genetic testing and screening are no different. This population of patients is children who cannot understand the purposes and outcomes of genetic screens and tests and adults afflicted with a mental or learning disability. For these populations, gaining informed consent is impossible [199]. In this respect, ethical consultation can play a significant role in aiding decision-making processes. Ethics consultants provide a forum for an open conversation that may dictate the importance of not presuming a patient's inability to communicate.

Instead, ethics consultation meetings may perpetuate patients' need for thorough capacity assessments before surrogacy or a similar substitutive method enacts. Patients deemed to lack capacity or cannot communicate effectively require specific attention due to their vulnerability [200]. A complex population includes patients who cannot communicate or give consent and may not benefit from a test or screen. However, the screen or test performed on this specific individual may very well aid another family member's diagnostic efforts that share the same genomic line. According to the Joint Committee on Medical Genetics, individuals who have hindered communication or decisional capacities should not be less altruistic than other patients. They should thus partake in procedures as long as they do not inflict harm to the agent [201].

Another primary issue regarding decision-making capacity or inhibited communication involves children who may not understand a genetic test or screen's implications. While studies indicate that testing and screening may occur in the child's best interest, the child should still be informed in a way that may perpetuate their understanding [202].Due to these illnesses' scope and nature, performing tests and screens for children susceptible to adult-onset diseases become more problematic. Rather than implementing a test or screen while the child is still unaware of their circumstances, many decisions regarding testing and screening for adult-onset genetic abnormalities defer until they can fully grasp the importance of their medical situation. In this respect, the child's autonomy comes to fruition as they develop cognitively. Conversely, many parents attempt to make these decisions for their child out of best interest standards, regardless of the test results or if the screen will have immediate results [203].

Although genetic testing and screening issues have their ethical roots in philosophical justification, it is essential to note the practicality and relevance of the information provided to families and patients during clinical ethics consultation. While it is important not to avoid the philosophical implications of genetic testing, screening, and modern technologies, families and patients must receive options and understand the impact of their choices.

# References

1. Pellegrino, Edmund D., and David C. Thomasma. "A Philosophical Method," In *A Philosophical Basis of Medical Practice: Toward a Philosophy and Ethic of the Healing Professions*, (New York: Oxford University Press, 1981), 39.
2. Pellegrino, Edmund D., and David C. Thomasma. "A Philosophical Method," In A Philosophical Basis of Medical Practice: *Toward a Philosophy and Ethic of the Healing Professions*, (New York: Oxford University Press, 1981), 39; cf. Pellegrino, E. D. "Philosophy of Medicine: Problematic and Potential." *Journal of Medicine and Philosophy* 1, no. 1 (1976): 5–31.

3. Pellegrino, Edmund D., and David C. Thomasma. "A Philosophical Method," In A Philosophical Basis of Medical Practice: *Toward a Philosophy and Ethic of the Healing Professions*, (New York: Oxford University Press, 1981), 39–43.

4. Pellegrino, Edmund D., and David C. Thomasma. "A Philosophical Method," In A Philosophical Basis of Medical Practice: *Toward a Philosophy and Ethic of the Healing Professions*, (New York: Oxford University Press, 1981), 39–43; *See* Marcum, James A. *An introductory philosophy of medicine: humanizing modern medicine.* (Dordrecht: Springer, 2010), 1–14.

5. Pellegrino, Edmund D., and David C. Thomasma. "A Philosophical Method," In A Philosophical Basis of Medical Practice: *Toward a Philosophy and Ethic of the Healing Professions*, (New York: Oxford University Press, 1981), 40–45.

6. Pellegrino, Edmund D., and David C. Thomasma. "A Philosophical Method," In A Philosophical Basis of Medical Practice: *Toward a Philosophy and Ethic of the Healing Professions*, (New York: Oxford University Press, 1981), 48–56.

7. Sugarman, Jeremy "Philosophy: Ethical Principles and Common Morality," In *Methods in Medical Ethics*, (Washington, D.C.: Georgetown University Press, 2010), 37–40.

8. Sugarman, Jeremy "Philosophy: Ethical Principles and Common Morality," *In Methods in Medical Ethics*, (Washington, D.C.: Georgetown University Press, 2010), 39.

9. Sugarman, Jeremy "Philosophy: Ethical Principles and Common Morality," In Methods in Medical Ethics, (Washington, D.C.: Georgetown University Press, 2010), 39.

10. Sugarman, Jeremy "Philosophy: Ethical Principles and Common Morality," In Methods in Medical Ethics, (Washington, D.C.: Georgetown University Press, 2010), 40; *See* Gert, Bernard. *Common morality: deciding what to do*, (New York: Oxford University Press, 2007), 27.

11. Sugarman, Jeremy "Philosophy: Ethical Principles and Common Morality," In Methods in Medical Ethics, (Washington, D.C.: Georgetown University Press, 2010), 40.

12. Sugarman, Jeremy "Philosophy: Ethical Principles and Common Morality," In Methods in Medical Ethics, (Washington, D.C.: Georgetown University Press, 2010), 40–41.

13. Sugarman, Jeremy "Philosophy: Ethical Principles and Common Morality," In Methods in Medical Ethics, (Washington, D.C.: Georgetown University Press, 2010), 40–41.

14. Sugarman, Jeremy "Philosophy: Ethical Principles and Common Morality," In Methods in Medical Ethics, (Washington, D.C.: Georgetown University Press, 2010), 39.

15. Sugarman, Jeremy "Philosophy: Ethical Principles and Common Morality," In Methods in Medical Ethics, (Washington, D.C.: Georgetown University Press, 2010), 40–41.

16. Sugarman, Jeremy "Philosophy: Ethical Principles and Common Morality," In Methods in Medical Ethics, (Washington, D.C.: Georgetown University Press, 2010), 42–43.

17. Ursin, Lars Øystein. "Personal Autonomy and Informed Consent." *Medicine, Health Care and Philosophy Med Health Care and Philos*, (2008): 17.

18. Ursin, Lars Øystein. "Personal Autonomy and Informed Consent." *Medicine, Health Care and Philosophy Med Health Care and Philos*, (2008): 18.

19. Ursin, Lars Øystein. "Personal Autonomy and Informed Consent." *Medicine, Health Care and Philosophy Med Health Care and Philos*, (2008): 18; *See* Brody, Baruch A., *Moral theory and moral judgments in medical ethics*, (Dordrecht: Kluwer Academic Pubs, 1988), 4–18. *See* Herman, Barbara, *Moral literacy*, (Cambridge, MA: Harvard University Press, 2008), 1–7.

20. Ursin, Lars Øystein. "Personal Autonomy and Informed Consent." *Medicine, Health Care and Philosophy Med Health Care and Philos*, (2008): 18.

21. Ursin, Lars Øystein. "Personal Autonomy and Informed Consent." *Medicine, Health Care and Philosophy Med Health Care and Philos*, (2008): 18.

22. Ursin, Lars Øystein. "Personal Autonomy and Informed Consent." *Medicine, Health Care and Philosophy Med Health Care and Philos*, (2008): 18–20; *See* Schmidt, Volker. "Patient Autonomy and the Ethics of Responsibility." *Asian Journal of Social Science* 36, no. 1 (2008): 148–49.

23. Beauchamp, Tom L., and James F. Childress. *Principles of Biomedical Ethics*. 5th ed. (New York, N.Y: Oxford University Press, 2013), 105–106.
24. Beauchamp, Tom L., and James F. Childress. *Principles of Biomedical Ethics*. 5th ed. (New York, N.Y: Oxford University Press, 2013), 105–106.
25. Martin, Rex. "Socrates on Disobedience to Law." *The Review of Metaphysics* 24, no. 1 (1970): 21–38.
26. Beauchamp, Tom L., and James F. Childress. *Principles of Biomedical Ethics*. 5th ed. (New York, N.Y: Oxford University Press, 2013), 105–106.
27. Beauchamp, Tom L., and James F. Childress. *Principles of Biomedical Ethics*. 5th ed. (New York, N.Y: Oxford University Press, 2013), 106–107.
28. Beauchamp, Tom L., and James F. Childress. *Principles of Biomedical Ethics*. 5th ed. (New York, N.Y: Oxford University Press, 2013), 106–107.
29. Beauchamp, Tom L., and James F. Childress. *Principles of Biomedical Ethics*. 5th ed. (New York, N.Y: Oxford University Press, 2013), 105–106.
30. Beauchamp, Tom L., and James F. Childress. *Principles of Biomedical Ethics*. 5th ed. (New York, N.Y: Oxford University Press, 2013), 107.
31. Pellegrino, Edmund D., and David C. Thomasma. *A Philosophical Basis of Medical Practice: Toward a Philosophy and Ethic of the Healing Professions*. (New York: Oxford University Press, 1981), 100–101.
32. Pellegrino, Edmund D., and David C. Thomasma. *A Philosophical Basis of Medical Practice: Toward a Philosophy and Ethic of the Healing Professions*. (New York: Oxford University Press, 1981), 22-11; cf. Longrigg, James. *Greek Rational Medicine: Philosophy and Medicine from Alcmaeon to the Alexandrians*, (Hoboken: Taylor and Francis, 2013), 4–25.
33. Pellegrino, Edmund D., and David C. Thomasma. "Ontology of the Body." In *A Philosophical Basis of Medical Practice: Toward a Philosophy and Ethic of the Healing Professions*. New York: Oxford University Press, (1981): 101.
34. Pellegrino, Edmund D., and David C. Thomasma. "Ontology of the Body." In *A Philosophical Basis of Medical Practice: Toward a Philosophy and Ethic of the Healing Professions*. New York: Oxford University Press, (1981): 101.
35. Pellegrino, Edmund D., and David C. Thomasma. "Ontology of the Body." In *A Philosophical Basis of Medical Practice: Toward a Philosophy and Ethic of the Healing Professions*. New York: Oxford University Press, (1981): 104; cf. Pellegrino, Edmund D. "The Internal Morality of Clinical Medicine: A Paradigm for the Ethics of the Helping and Healing Professions." *The Journal of Medicine and Philosophy* 26, no. 6 (2001): 559–79.
36. Pellegrino, Edmund D., and David C. Thomasma. "Ontology of the Body." In *A Philosophical Basis of Medical Practice: Toward a Philosophy and Ethic of the Healing Professions*. New York: Oxford University Press, (1981): 117.
37. Pellegrino, Edmund D., and David C. Thomasma. "Ontology of the Body." In *A Philosophical Basis of Medical Practice: Toward a Philosophy and Ethic of the Healing Professions*. New York: Oxford University Press, (1981): 117.
38. Pellegrino, Edmund D., and David C. Thomasma. "Ontology of the Body." In *A Philosophical Basis of Medical Practice: Toward a Philosophy and Ethic of the Healing Professions*. New York: Oxford University Press, (1981): 117–18.
39. Beauchamp, Tom L. "Informed Consent: Its History, Meaning, and Present Challenges." *Cambridge Quarterly of Healthcare Ethics Cambridge Healthcare Ethics Quarterly*, 20 no. 4, (2011): 515.
40. Beauchamp, Tom L. "Informed Consent: Its History, Meaning, and Present Challenges." *Cambridge Quarterly of Healthcare Ethics Cambridge Healthcare Ethics Quarterly*, 20 no. 4, (2011): 515.
41. Beauchamp, Tom L. "Informed Consent: Its History, Meaning, and Present Challenges." *Cambridge Quarterly of Healthcare Ethics Cambridge Healthcare Ethics Quarterly*, 20 no. 4, (2011): 515; cf. Berg, Jessica W., *Informed consent: legal theory and clinical practice*. (Oxford: Oxford University Press, 2007): 4–14.

42. Beauchamp, Tom L. "Informed Consent: Its History, Meaning, and Present Challenges." *Cambridge Quarterly of Healthcare Ethics Cambridge Healthcare Ethics Quarterly*, 20 no. 4, (2011): 515–16.
43. Beauchamp, Tom L. "Informed Consent: Its History, Meaning, and Present Challenges." *Cambridge Quarterly of Healthcare Ethics Cambridge Healthcare Ethics Quarterly*, 20 no. 4, (2011): 515–16.
44. Beauchamp, Tom L. "Informed Consent: Its History, Meaning, and Present Challenges." *Cambridge Quarterly of Healthcare Ethics Cambridge Healthcare Ethics Quarterly*, 20 no. 4, (2011): 515–23.
45. Beauchamp, Tom L. "Informed Consent: Its History, Meaning, and Present Challenges." *Cambridge Quarterly of Healthcare Ethics Cambridge Healthcare Ethics Quarterly*, 20 no. 4, (2011): 516.
46. Beauchamp, Tom L. "Informed Consent: Its History, Meaning, and Present Challenges." *Cambridge Quarterly of Healthcare Ethics Cambridge Healthcare Ethics Quarterly*, 20 no. 4, (2011): 516.
47. Beauchamp, Tom L. "Informed Consent: Its History, Meaning, and Present Challenges." *Cambridge Quarterly of Healthcare Ethics Cambridge Healthcare Ethics Quarterly*, 20 no. 4, (2011): 516–17.
48. Beauchamp, Tom L. "Informed Consent: Its History, Meaning, and Present Challenges." *Cambridge Quarterly of Healthcare Ethics Cambridge Healthcare Ethics Quarterly*, 20 no. 4, (2011): 517–18; *See* Manson, Neil C., and Onora ONeill, *Rethinking informed consent in bioethics*, (Cambridge: Cambridge University Press, 2008): 1–7.
49. Beauchamp, Tom L. "Informed Consent: Its History, Meaning, and Present Challenges." *Cambridge Quarterly of Healthcare Ethics Cambridge Healthcare Ethics Quarterly*, 20 no. 4, (2011): 517–18.
50. Beauchamp, Tom L. "Informed Consent: Its History, Meaning, and Present Challenges." *Cambridge Quarterly of Healthcare Ethics Cambridge Healthcare Ethics Quarterly*, 20 no. 4, (2011): 517–18.
51. Beauchamp, Tom L. "Informed Consent: Its History, Meaning, and Present Challenges." *Cambridge Quarterly of Healthcare Ethics Cambridge Healthcare Ethics Quarterly*, 20 no. 4, (2011): 517–18.
52. Beauchamp, Tom L. "Informed Consent: Its History, Meaning, and Present Challenges." *Cambridge Quarterly of Healthcare Ethics Cambridge Healthcare Ethics Quarterly*, 20 no. 4, (2011): 518.
53. Beauchamp, Tom L. "Informed Consent: Its History, Meaning, and Present Challenges." *Cambridge Quarterly of Healthcare Ethics Cambridge Healthcare Ethics Quarterly*, 20 no. 4, (2011): 518.
54. Alderson, Priscilla, and Virginia Morrow, *The Ethics of Research with Children and Young People: A Practical Handbook,* 2nd ed., (London: SAGE, 2011), 100.
55. Alderson, Priscilla, and Virginia Morrow, *The Ethics of Research with Children and Young People: A Practical Handbook,* 2nd ed., (London: SAGE, 2011), 100.*See* Greenhalgh, Trisha. *Narrative based medicine: dialogue and discourse in clinical practice.* (London: BMJ, 2004), 4–45.
56. Alderson, Priscilla, and Virginia Morrow, *The Ethics of Research with Children and Young People: A Practical Handbook,* 2nd ed., (London: SAGE, 2011), 85. *See* Greenhalgh, Trisha. *Narrative based medicine: dialogue and discourse in clinical practice.* (London: BMJ, 2004), 4–45.
57. Alderson, Priscilla, and Virginia Morrow, *The Ethics of Research with Children and Young People: A Practical Handbook,* 2nd ed., (London: SAGE, 2011), 85. *See* Greenhalgh, Trisha. *Narrative based medicine: dialogue and discourse in clinical practice.* (London: BMJ, 2004), 4–45.
58. Alderson, Priscilla, and Virginia Morrow, *The Ethics of Research with Children and Young People: A Practical Handbook,* 2nd ed., (London: SAGE, 2011), 85; *See* Greenhalgh, Trisha.

*Narrative based medicine: dialogue and discourse in clinical practice.* (London: BMJ, 2004), 4–45.

59. Alderson, Priscilla, and Virginia Morrow, *The Ethics of Research with Children and Young People: A Practical Handbook,* 2nd ed., (London: SAGE, 2011), 85.

60. Alderson, Priscilla, and Virginia Morrow, *The Ethics of Research with Children and Young People: A Practical Handbook,* 2nd ed., (London: SAGE, 2011), 85.

61. Alderson, Priscilla, and Virginia Morrow, *The Ethics of Research with Children and Young People: A Practical Handbook,* 2nd ed., (London: SAGE, 2011), 86.

62. Alderson, Priscilla, and Virginia Morrow, *The Ethics of Research with Children and Young People: A Practical Handbook,* 2nd ed., (London: SAGE, 2011), 85.

63. Alderson, Priscilla, and Virginia Morrow, *The Ethics of Research with Children and Young People: A Practical Handbook,* 2nd ed., (London: SAGE, 2011), 85; cf. Charles, Cathy, Amiram Gafni, and Tim Whelan, "Decision-making in the physician–patient encounter: revisiting the shared treatment decision-making model," *Social Science & Medicine* 49, no. 5 (1999): 651–61.

64. Gambetta, Diego, *Trust: Making and Breaking Cooperative Relations,* (New York: Blackwell, 2000), 49–72.

65. Chadwick, Ruth F., Henk Ten Have, and Eric M. Meslin. "Informed Consent." In *The SAGE Handbook of Health Care Ethics,* (Los Angeles, California: SAGE, 2011), 107.

66. Chadwick, Ruth F., Henk Ten Have, and Eric M. Meslin. "Informed Consent." In *The SAGE Handbook of Health Care Ethics,* (Los Angeles, California: SAGE, 2011), 107; *See* Kuhse, Helga, and Peter Singer. *A companion to bioethics,* (Chichester, U.K.: Wiley-Blackwell, 2012), 32–34.

67. Chadwick, Ruth F., Henk Ten Have, and Eric M. Meslin. "Informed Consent." In *The SAGE Handbook of Health Care Ethics,* (Los Angeles, California: SAGE, 2011), 108.

68. Chadwick, Ruth F., Henk Ten Have, and Eric M. Meslin. "Informed Consent." In *The SAGE Handbook of Health Care Ethics,* (Los Angeles, California: SAGE, 2011), 108.

69. Chadwick, Ruth F., Henk Ten Have, and Eric M. Meslin. "Informed Consent." In *The SAGE Handbook of Health Care Ethics,* (Los Angeles, California: SAGE, 2011), 108.

70. Chadwick, Ruth F., Henk Ten Have, and Eric M. Meslin. "Informed Consent." In *The SAGE Handbook of Health Care Ethics,* (Los Angeles, California: SAGE, 2011), 108.

71. Emanuel, Ezekiel J., Christine C. Grady, Robert A. Crouch, Reidar K. Lie, Franklin G. Miller, and DavidD. Wendler, eds, "Philosophical Justifications of Informed Consent in Research" In *The Oxford textbook of clinical research ethics,* (Oxford University Press, 2008), 607.

72. Emanuel, Ezekiel J., Christine C. Grady, Robert A. Crouch, Reidar K. Lie, Franklin G. Miller, and DavidD. Wendler, eds, "Philosophical Justifications of Informed Consent in Research" In *The Oxford textbook of clinical research ethics,* (Oxford University Press, 2008), 607; *See* Roberts, Laura Weiss, "Informed Consent and the Capacity for Voluntarism," 1, no. 4, (2003): 407–414.

73. Oliver, Paul. "Research and the Respondent: Ethical Issues Before the Research Commences." In *The Student's Guide to Research Ethics.* 2nd ed. Maidenhead, Berkshire, England: McGraw-Hill/Open University Press, (2010): 26–28.

74. Oliver, Paul. "Research and the Respondent: Ethical Issues Before the Research Commences." In *The Student's Guide to Research Ethics.* 2nd ed. Maidenhead, Berkshire, England: McGraw-Hill/Open University Press, (2010): 26–28.

75. Emanuel, Ezekiel J., Christine C. Grady, Robert A. Crouch, Reidar K. Lie, Franklin G. Miller, and DavidD. Wendler, eds, "Exploitation in Clinical Research," In *The Oxford textbook of clinical research ethics,* (Oxford University Press, 2008), 201; cf, Krupp, Brandon H. "Professional Boundaries: Safeguarding the Physician–Patient Relationship." (*Dermatoethics,* 2011), 61–66.

76. Emanuel, Ezekiel J., Christine C. Grady, Robert A. Crouch, Reidar K. Lie, Franklin G. Miller, and DavidD. Wendler, eds, "Exploitation in Clinical Research," In *The Oxford textbook of clinical research ethics,* (Oxford University Press, 2008), 201.

77. Emanuel, Ezekiel J., Christine C. Grady, Robert A. Crouch, Reidar K. Lie, Franklin G. Miller, and DavidD. Wendler, eds, "Exploitation in Clinical Research," In *The Oxford textbook of clinical research ethics*, (Oxford University Press, 2008), 202.

78. Emanuel, Ezekiel J., Christine C. Grady, Robert A. Crouch, Reidar K. Lie, Franklin G. Miller, and DavidD. Wendler, eds, "Exploitation in Clinical Research," In *The Oxford textbook of clinical research ethics*, (Oxford University Press, 2008), 203; *See* Cameron, Lois, and Joan Murphy, "Obtaining consent to participate in research: the issues involved in including people with a range of learning and communication disabilities," *British Journal of Learning Disabilities* 35, no. 2 (2007): 113–20.

79. Emanuel, Ezekiel J., Christine C. Grady, Robert A. Crouch, Reidar K. Lie, Franklin G. Miller, and David D. Wendler, eds, "Exploitation in Clinical Research," In *The Oxford textbook of clinical research ethics*, (Oxford University Press, 2008), 203–204.

80. Beauchamp, Tom L., and James F. Childress. *Principles of Biomedical Ethics*. 5th ed. (New York, N.Y: Oxford University Press, 2013), 267.

81. Beauchamp, Tom L., and James F. Childress. *Principles of Biomedical Ethics*. 5th ed. (New York, N.Y: Oxford University Press, 2013), 267.

82. Beauchamp, Tom L., and James F. Childress. *Principles of Biomedical Ethics*. 5th ed. (New York, N.Y: Oxford University Press, 2013), 267–68.

83. Gert, Bernard, and Charles M. Culver, "Adequate Information, Competence, and Coercion," In *Bioethics a Systematic Approach*, 2nd ed., (Oxford: Oxford University Press, 2006), 232–234; *See* Dugdale, David C., Ronald Epstein, and Steven Z. Pantilat, "Time and the patient-physician relationship," *Journal of General Internal Medicine* 14, no. S1 (1999): 2–19.

84. Beauchamp, Tom L., and James F. Childress. *Principles of Biomedical Ethics*. 5th ed. (New York, N.Y: Oxford University Press, 2013), 267–68.

85. Gert, Bernard, and Charles M. Culver, "Adequate Information, Competence, and Coercion," In *Bioethics a Systematic Approach*, 2nd ed., (Oxford: Oxford University Press, 2006), 232.

86. Gert, Bernard, and Charles M. Culver, "Adequate Information, Competence, and Coercion," In *Bioethics a Systematic Approach*, 2nd ed., (Oxford: Oxford University Press, 2006), 232.

87. Gert, Bernard, and Charles M. Culver, "Adequate Information, Competence, and Coercion," In *Bioethics a Systematic Approach*, 2nd ed., (Oxford: Oxford University Press, 2006), 232.

88. Gert, Bernard, and Charles M. Culver, "Adequate Information, Competence, and Coercion," In *Bioethics a Systematic Approach*, 2nd ed., (Oxford: Oxford University Press, 2006), 232; *See* Trotter, Griffin, *The Ethics of Coercion in Mass Casualty Medicine*, (Baltimore: Johns Hopkins University Press, 2009), 24–85.

89. Gert, Bernard, and Charles M. Culver, "Adequate Information, Competence, and Coercion," In *Bioethics a Systematic Approach*, 2nd ed., (Oxford: Oxford University Press, 2006), 232.

90. Gert, Bernard, and Charles M. Culver, "Adequate Information, Competence, and Coercion," In *Bioethics a Systematic Approach*, 2nd ed., (Oxford: Oxford University Press, 2006), 232–34.

91. Gert, Bernard, and Charles M. Culver, "Adequate Information, Competence, and Coercion," In *Bioethics a Systematic Approach*, 2nd ed., (Oxford: Oxford University Press, 2006), 232; *See* Pellegrino, Edmund D. "Autonomy and coercion in disease prevention and health promotion," *Theoretical Medicine* 5, no. 1 (1984): 83–91.

92. Gert, Bernard, and Charles M. Culver, "Adequate Information, Competence, and Coercion," In *Bioethics a Systematic Approach*, 2nd ed., (Oxford: Oxford University Press, 2006), 237.

93. Gert, Bernard, and Charles M. Culver, "Adequate Information, Competence, and Coercion," In *Bioethics a Systematic Approach*, 2nd ed., (Oxford: Oxford University Press, 2006), 237; cf. Binder, Martin, and Leonhard K. Lades, "Autonomy-Enhancing Paternalism," *Levy Economics Institute, Working Papers Series*, no. 800, (2014): 1–13.

94. Derr, Susan Dawson. "Hospital Ethics Committees: Historical Development, Current Issues, and Recommendations." Ph.D. diss., (Drew University, 2009), 1–15.

95. Gert, Bernard, and Charles M. Culver, "Adequate Information, Competence, and Coercion," In *Bioethics a Systematic Approach*, 2nd ed., (Oxford: Oxford University Press, 2006), 237.

96. Gert, Bernard, and Charles M. Culver, "Adequate Information, Competence, and Coercion," In *Bioethics a Systematic Approach*, 2nd ed., (Oxford: Oxford University Press, 2006), 237. *See* Vandeveer, Donald. *Paternalistic Intervention: The Moral Bounds on Benevolence.* (Princeton University Press, 2014), 45–50.

97. Gert, Bernard, and Charles M. Culver, "Adequate Information, Competence, and Coercion," In *Bioethics a Systematic Approach*, 2nd ed., (Oxford: Oxford University Press, 2006), 237; *See* Hauser, Marc, Fiery Cushman, Liane Young, R. Kang-Xing Jin, and John Mikhail. "A Dissociation Between Moral Judgments and Justifications," *Mind & Language* 22, no. 1 (2007): 1–21.

98. Gert, Bernard, and Charles M. Culver, "Adequate Information, Competence, and Coercion," In *Bioethics a Systematic Approach*, 2nd ed., (Oxford: Oxford University Press, 2006), 237–238.

99. Gert, Bernard, and Charles M. Culver, "Adequate Information, Competence, and Coercion," In *Bioethics a Systematic Approach*, 2nd ed., (Oxford: Oxford University Press, 2006), 238.

100. Gert, Bernard, and Charles M. Culver, "Adequate Information, Competence, and Coercion," In *Bioethics a Systematic Approach*, 2nd ed., (Oxford: Oxford University Press, 2006), 238; *See* Hauser, Marc, Fiery Cushman, Liane Young, R. Kang-Xing Jin, and John Mikhail. "A Dissociation Between Moral Judgments and Justifications." *Mind & Language* 22, no. 1 (2007): 1–21.

101. Gert, Bernard, and Charles M. Culver, "Adequate Information, Competence, and Coercion," In *Bioethics a Systematic Approach*, 2nd ed., (Oxford: Oxford University Press, 2006), 236–240.

102. Gert, Bernard, and Charles M. Culver, "Adequate Information, Competence, and Coercion," In *Bioethics a Systematic Approach*, 2nd ed., (Oxford: Oxford University Press, 2006), 240–245.

103. Gert, Bernard, and Charles M. Culver, "Adequate Information, Competence, and Coercion," In *Bioethics a Systematic Approach*, 2nd ed., (Oxford: Oxford University Press, 2006), 246–248.

104. Gert, Bernard, and Charles M. Culver, "Adequate Information, Competence, and Coercion," In *Bioethics a Systematic Approach*, 2nd ed., (Oxford: Oxford University Press, 2006), 248–252; *See* Tomlinson, T., and H. Brody. "Ethics and communication in do-not-resuscitate orders," *Dimensions of Critical Care Nursing* 7, no. 4 (1988): 226.

105. Beauchamp, Tom L., and James F. Childress. "Justice." In *Principles of Biomedical Ethics*. 5th ed. (New York, N.Y.: Oxford University Press, 2001): 214–15.

106. Beauchamp, Tom L., and James F. Childress. "Justice." In *Principles of Biomedical Ethics*. 5th ed. (New York, N.Y.: Oxford University Press, 2001): 214–15.

107. Beauchamp, Tom L., and James F. Childress. "Justice." In *Principles of Biomedical Ethics*. 5th ed. (New York, N.Y.: Oxford University Press, 2001): 215.

108. Beauchamp, Tom L., and James F. Childress. "Justice." In *Principles of Biomedical Ethics*. 5th ed. (New York, N.Y.: Oxford University Press, (2001): 214–15.

109. Beauchamp, Tom L., and James F. Childress. "Justice." In *Principles of Biomedical Ethics*. 5th ed. (New York, N.Y.: Oxford University Press, 2001): 216.

110. Beauchamp, Tom L., and James F. Childress. "Justice." In *Principles of Biomedical Ethics*. 5th ed. (New York, N.Y.: Oxford University Press, 2001): 216–17.

111. Beauchamp, Tom L., and James F. Childress. "Justice." In *Principles of Biomedical Ethics*. 5th ed. (New York, N.Y.: Oxford University Press, 2001). 217.

112. Bernstein, Mark, and Kerry Bowman, "Should a Medical/Surgical Specialist with Formal Training in Bioethics Provide Health Care Ethics Consultation in his/her Own Area of Specialty?." In *HEC Forum*, vol. 15, no. 3, Kluwer Academic Publishers, (2003): 274."

113. Finder, Stuart G. "Is Consent Necessary for Ethics Consultation?" *Cambridge Quarterly of Healthcare Ethics Camb Q Healthc Ethics*, (2009): 384.

114. Finder, Stuart G. "Is Consent Necessary for Ethics Consultation?" *Cambridge Quarterly of Healthcare Ethics Camb Quarterly Healthcare Ethics*, (2009): 384; cf, Savulescu, J.,

I. Chalmers, and J. Blunt. "Are research ethics committees behaving unethically? Some suggestions for improving performance and accountability." *Bmj* 313, no. 7069 (1996): 1390–393.

115. Finder, Stuart G. "Is Consent Necessary for Ethics Consultation?" *Cambridge Quarterly of Healthcare Ethics Camb Q Healthc Ethics*, (2009): 384.

116. Finder, Stuart G. "Is Consent Necessary for Ethics Consultation?" *Cambridge Quarterly of Healthcare Ethics Camb Q Healthc Ethics*, (2009): 384. *See* Stevens, M. L. Tina. "The Quinlan Case Revisited: A History of the Cultural Politics of Medicine and the Law." *Journal of Health Politics, Policy and Law* 21, no. 2 (1996): 347–66.f advocates that would advse and aid them during the course of their treatment. es aided patients by granting them a pannel ees

117. Bernstein, Mark, and Kerry Bowman, "Should a Medical/Surgical Specialist with Formal Training in Bioethics Provide Health Care Ethics Consultation in his/her Own Area of Specialty?." In *HEC Forum*, vol. 15, no. 3, Kluwer Academic Publishers, (2003): 274.

118. Bernstein, Mark, and Kerry Bowman, "Should a Medical/Surgical Specialist with Formal Training in Bioethics Provide Health Care Ethics Consultation in his/her Own Area of Specialty?." In *HEC Forum*, vol. 15, no. 3, Kluwer Academic Publishers, (2003): 274.

119. Bernstein, Mark, and Kerry Bowman, "Should a Medical/Surgical Specialist with Formal Training in Bioethics Provide Health Care Ethics Consultation in his/her Own Area of Specialty?." In *HEC Forum*, vol. 15, no. 3, Kluwer Academic Publishers, (2003): 274.

120. Bernstein, Mark, and Kerry Bowman, "Should a Medical/Surgical Specialist with Formal Training in Bioethics Provide Health Care Ethics Consultation in his/her Own Area of Specialty?." In *HEC Forum*, vol. 15, no. 3, Kluwer Academic Publishers, (2003): 281.

121. Bernstein, Mark, and Kerry Bowman, "Should a Medical/Surgical Specialist with Formal Training in Bioethics Provide Health Care Ethics Consultation in his/her Own Area of Specialty?." In *HEC Forum*, vol. 15, no. 3, Kluwer Academic Publishers, (2003): 281–282; cf. Stempsey, William E. "Philosophy of Medicine Is What Philosophers of Medicine Do." *Perspectives in Biology and Medicine* 51, no. 3 (2008): 379–91.

122. Bernstein, Mark, and Kerry Bowman, "Should a Medical/Surgical Specialist with Formal Training in Bioethics Provide Health Care Ethics Consultation in his/her Own Area of Specialty?." In *HEC Forum*, vol. 15, no. 3, Kluwer Academic Publishers, (2003): 281–282.

123. Oliver, Paul. "Conclusion: The Role of The Researcher." In *The Student's Guide to Research Ethics*. 2nd ed. Maidenhead, (Berkshire, England: McGraw-Hill/Open University Press, 2010), 171–175.

124. Fiscella, Kevin, Jonathan N. Tobin, Jennifer K. Carroll, Hua He, and Gbenga Ogedegbe. "Ethical oversight in quality improvement and quality improvement research: new approaches to promote a learning health care system," *BMC Ethics* 16, no. 63 (2015): 1–3.

125. Fiscella, Kevin, Jonathan N. Tobin, Jennifer K. Carroll, Hua He, and Gbenga Ogedegbe. "Ethical oversight in quality improvement and quality improvement research: new approaches to promote a learning health care system," *BMC Ethics* 16, no. 63 (2015): 3–6. *See* Asher, Shellie, "Stewardship of Health Care Resources," In *Ethical Dilemmas in Emergency Medicine*, (Ohio: Cambridge University Press 2015), 266–77.

126. Oliver, Paul. "Conclusion: The Role of The Researcher." In *The Student's Guide to Research Ethics*. 2nd ed. Maidenhead, (Berkshire, England: McGraw-Hill/Open University Press, 2010), 171.

127. Oliver, Paul. "Conclusion: The Role of The Researcher." In *The Student's Guide to Research Ethics*. 2nd ed. Maidenhead, (Berkshire, England: McGraw-Hill/Open University Press, 2010), 171.

128. Oliver, Paul. "Conclusion: The Role of The Researcher." In *The Student's Guide to Research Ethics*. 2nd ed. Maidenhead, (Berkshire, England: McGraw-Hill/Open University Press, 2010), *See* Bowie, Norman E. *Business ethics: a Kantian perspective*, (New York, NY: Cambridge University Press, 2017), 11–24.

129. Oliver, Paul. "Conclusion: The Role of The Researcher." In *The Student's Guide to Research Ethics*. 2nd ed. Maidenhead, (Berkshire, England: McGraw-Hill/Open University Press,

2010), 172. *See* Sherman, Nancy. *Making a necessity of virtue: Aristotle and Kant on virtue.* (Cambridge: Cambridge University Press, 2004), 3–13.

130. Oliver, Paul. "Conclusion: The Role of The Researcher." In *The Student's Guide to Research Ethics.* 2nd ed. Maidenhead, (Berkshire, England: McGraw-Hill/Open University Press, 2010), 172.

131. Oliver, Paul. "Conclusion: The Role of The Researcher." In *The Student's Guide to Research Ethics.* 2nd ed. Maidenhead, (Berkshire, England: McGraw-Hill/Open University Press, 2010), 172.

132. Bernstein, Mark, and Kerry Bowman, "Should a Medical/Surgical Specialist with Formal Training in Bioethics Provide Health Care Ethics Consultation in his/her Own Area of Specialty?." In *HEC Forum*, vol. 15, no. 3, Kluwer Academic Publishers, (2003): 282.

133. Chadwick, Ruth F., Henk Ten Have, and Eric M. Meslin. "Research Ethics," In *The SAGE Handbook of Health Care Ethics.* (Los Angeles, California: SAGE, 2011), 323.

134. Bernstein, Mark, and Kerry Bowman, "Should a Medical/Surgical Specialist with Formal Training in Bioethics Provide Health Care Ethics Consultation in his/her Own Area of Specialty?." In *HEC Forum*, vol. 15, no. 3, Kluwer Academic Publishers, (2003): 282.

135. Bernstein, Mark, and Kerry Bowman, "Should a Medical/Surgical Specialist with Formal Training in Bioethics Provide Health Care Ethics Consultation in his/her Own Area of Specialty?." In *HEC Forum*, vol. 15, no. 3, Kluwer Academic Publishers, (2003): 282.

136. Schaefer, Bradley G., and James N. Thompson, Jr. *Medical Genetics: An Integrated Approach.* (McGraw-Hill Education, 2014), 9.

137. Schaefer, Bradley G., and James N. Thompson, Jr. *Medical Genetics: An Integrated Approach.* (McGraw-Hill Education, 2014), 9–10.

138. Schaefer, Bradley G., and James N. Thompson, Jr. *Medical Genetics: An Integrated Approach.* (McGraw-Hill Education, 2014), 9–10. *See* McElheny, Victor K. *Drawing the map of life: inside the Human Genome Project,* (New York, NY: Basic Books, a member of the Perseus Books Group, 2012) 12–45.

139. Schaefer, Bradley G., and James N. Thompson, Jr. *Medical Genetics: An Integrated Approach.* (McGraw-Hill Education, 2014), 11.

140. Schaefer, Bradley G., and James N. Thompson, Jr. *Medical Genetics: An Integrated Approach.* (McGraw-Hill Education, 2014), 11.

141. Schaefer, Bradley G., and James N. Thompson, Jr. *Medical Genetics: An Integrated Approach.* (McGraw-Hill Education, 2014), 15.

142. Schaefer, Bradley G., and James N. Thompson, Jr. *Medical Genetics: An Integrated Approach.* (McGraw-Hill Education, 2014), 15.

143. Buchanan, Allen. *Beyond Humanity?* (New York, New York: Oxford University Press, 2011), 209–230.

144. Schaefer, Bradley G., and James N. Thompson, Jr. *Medical Genetics: An Integrated Approach.* (McGraw-Hill Education, 2014), 229–31.

145. Schaefer, Bradley G., and James N. Thompson, Jr. *Medical Genetics: An Integrated Approach.* (McGraw-Hill Education, 2014), 230.

146. Schaefer, Bradley G., and James N. Thompson, Jr. *Medical Genetics: An Integrated Approach.* (McGraw-Hill Education, 2014), 236. *See* Have, Henk AMJ. "Medical technology assessment and ethics." *Hastings Center Report* 25, no. 5 (1995): 13–19.

147. Fitzsimons, Peter John, "Biotechnology, Ethics and Education," *Studies in Philosophy and Education Studies in Philosophical Education,* 26, no. 1, (2006): 1–5.

148. Buchanan, Allen. *Beyond Humanity?* (New York, New York: Oxford University Press, 2011). 209–40.

149. Schaefer, Bradley G., and James N. Thompson, Jr. *Medical Genetics: An Integrated Approach.* (McGraw-Hill Education, 2014), 236.

150. Schaefer, Bradley G., and James N. Thompson, Jr. *Medical Genetics: An Integrated Approach.* (McGraw-Hill Education, 2014), 236–37.

151. Schaefer, Bradley G., and James N. Thompson, Jr. *Medical Genetics: An Integrated Approach.* (McGraw-Hill Education, 2014). 236.
152. Schaefer, Bradley G., and James N. Thompson, Jr. *Medical Genetics: An Integrated Approach.* (McGraw-Hill Education, 2014), 236–37.
153. Schaefer, Bradley G., and James N. Thompson, Jr. *Medical Genetics: An Integrated Approach.* (McGraw-Hill Education, 2014), 236–37. *See* Rothenberg, Karen H., and Elizabeth Jean. Thomson. *Women and prenatal testing: facing the challenges of genetic technology,* (Columbus: Ohio State University Press, 1994).
154. Schaefer, Bradley G., and James N. Thompson, Jr. *Medical Genetics: An Integrated Approach.* (McGraw-Hill Education, 2014), 236–37. *See* Eng, Charis, and Jan Vijg, "Genetic testing: The problems and the promise," *Nature Biotechnology* 15, no. 5 (1997): 422–26.
155. Schaefer, Bradley G., and James N. Thompson, Jr. *Medical Genetics: An Integrated Approach.* (McGraw-Hill Education, 2014). 244–45.
156. Schaefer, Bradley G., and James N. Thompson, Jr. *Medical Genetics: An Integrated Approach.* (McGraw-Hill Education, 2014), 246.
157. Pols, Jeannette. "Towards an Empirical Ethics in Care: Relations with Technologies in Health Care." *Medicine, Health Care and Philosophy Med Health Care and Philos,* (2014): 81–90.
158. Heidegger, Martin, and David Farrell Krell. "The Question Concerning Technology," In *Basic Writings: Martin Heindegger,* (London: Routledge, 2010), 341. *See* Verbeek, Peter-Paul, "Materializing Morality," *Science, Technology, & Human Values* 31, no. 3 (2006): 361–80.
159. Buchanan, Allen. *Beyond Humanity?* (New York, New York: Oxford University Press, 2011), 35–40.
160. Heidegger, Martin, and David Farrell Krell. "The Question Concerning Technology." In *Basic Writings: Martin Heindegger,* 307–341. (London: Routledge, 2010), 307 – 315.
161. Schalow, Frank. "The Gods and Technology: A Reading of Heidegger." *Journal of Phenomenological Psychology,* (2008): 71. *See* Bailey, Jesse I. "Enframing The Flesh: Heidegger, Transhumanism, And The Body As "Standing Reserve,'" *Journal of Evolution & Technology* No. 24.2 (2014): 44–62.
162. Schalow, Frank. "The Gods and Technology: A Reading of Heidegger." *Journal of Phenomenological Psychology,* (2008): 67–70.
163. Schalow, Frank, "The Gods and Technology: A Reading of Heidegger," *Journal of Phenomenological Psychology,* (2008): 71.
164. Heidegger, Martin, and David Farrell Krell, "The Question Concerning Technology," In *Basic Writings: Martin Heindegger,* (London: Routledge, 2010), 307–41.
165. Schalow, Frank. "The Gods and Technology: A Reading of Heidegger." *Journal of Phenomenological Psychology,* (2008) 71.
166. Heidegger, Martin, and David Farrell Krell. "The Question Concerning Technology." In *Basic Writings: Martin Heindegger,* (London: Routledge, 2010), 320–30.
167. Schwab, Abraham Paul. *The Human Microbiome: Ethical, Legal and Social Concerns.* Edited by Rosamond Rhodes and Nada Gligorov, (New York, New York: Oxford University Press, 2013), 66–69.
168. Schalow, Frank. "The Gods and Technology: A Reading of Heidegger." *Journal of Phenomenological Psychology,* (2008): 71.
169. Schalow, Frank. "The Gods and Technology: A Reading of Heidegger." *Journal of Phenomenological Psychology,* (200): 71–75. *See* Babich, Babette, "O, Superman! Or being Towards Transhumanism: Martin Heidegger, Günther Anders, and Media Aesthetics," *Maison des Sciences de l'Homme et de la Société (Sofia),* no. 36, (2013): 112–46.
170. Schalow, Frank. "The Gods and Technology: A Reading of Heidegger." *Journal of Phenomenological Psychology,* (2008): 75–76.
171. Schalow, Frank. "The Gods and Technology: A Reading of Heidegger." *Journal of Phenomenological Psychology,* (2008): 77.
172. Pols, Jeannette. "Towards an Empirical Ethics in Care: Relations with Technologies in Health Care." *Medicine, Health Care and Philosophy Med Health Care and Philos,* (2014) 81–90.

173. Schalow, Frank. "The Gods and Technology: A Reading of Heidegger." *Journal of Phenomenological Psychology*, (2008): 77–78.
174. Schalow, Frank. "The Gods and Technology: A Reading of Heidegger." *Journal of Phenomenological Psychology*, (2008): 142–50.
175. Schalow, Frank. "The Gods and Technology: A Reading of Heidegger." *Journal of Phenomenological Psychology*, (2008): 150–63.
176. Schalow, Frank. "The Gods and Technology: A Reading of Heidegger." *Journal of Phenomenological Psychology*, (2008): 160–68.
177. More, Max, and Natasha Vita-More. *The transhumanist reader: classical and contemporary essays on the science, technology, and philosophy of the human future*, (Malden: Wiley-Blackwell, J. Wiley & Sons, 2013), 1–8.
178. More, Max, and Natasha Vita-More. *The transhumanist reader: classical and contemporary essays on the science, technology, and philosophy of the human future*, (Malden: Wiley-Blackwell, J. Wiley & Sons, 2013), 8–12.
179. Schwab, Abraham Paul. *The Human Microbiome: Ethical, Legal and Social Concerns.* Edited by Rosamond Rhodes and Nada Gligorov. New York, New York: (Oxford University Press, 2013), 52–58.
180. More, Max, and Natasha Vita-More. *The transhumanist reader: classical and contemporary essays on the science, technology, and philosophy of the human future*, (Malden: Wiley-Blackwell, J. Wiley & Sons, 2013), 12–15.
181. More, Max, and Natasha Vita-More. *The transhumanist reader: classical and contemporary essays on the science, technology, and philosophy of the human future*, (Malden: Wiley-Blackwell, J. Wiley & Sons, 2013), 14–15.
182. Schalow, Frank. "The Gods and Technology: A Reading of Heidegger." *Journal of Phenomenological Psychology*, (2008): 175 - 200.
183. Schalow, Frank. "The Gods and Technology: A Reading of Heidegger." *Journal of Phenomenological Psychology*, (2008): 205–230.
184. Lo, Bernard. *Resolving Ethical Dilemmas: A Guide for Clinicians*. 4th ed. (Philadelphia: Wolters Kluwer Health/Lippincott Williams & Wilkins, 2009). 3 11.
185. Baily, Mary Ann, and Thomas H. Murray. "Ethics, Evidence, and Cost in Newborn Screening." *The Hastings Center Report*, May/June, 38, no. 3 (2008): 23–25.
186. Lo, Bernard. *Resolving Ethical Dilemmas: A Guide for Clinicians*. 4th ed. (Philadelphia: Wolters Kluwer Health/Lippincott Williams & Wilkins, 2009), 311–20.
187. Baily, Mary Ann, and Thomas H. Murray. "Ethics, Evidence, and Cost in Newborn Screening." *The Hastings Center Report*, May/June, 38, no. 3 (2008): 24–25.
188. Kennedy, Shelley, Beth K. Potter, Kumanan Wilson, Lawrence Fisher, Michael Geraghty, Jennifer Milburn, and Pranesh Chakraborty. "The First Three Years of Screening for Medium Chain Acyl-CoA Dehydrogenase Deficiency (MCADD) by Newborn Screening Ontario." *BMC Pediatrics BMC Pediatr* 10, no. 1 (2010): 82–90.
189. Baily, Mary Ann, and Thomas H. Murray. "Ethics, Evidence, and Cost in Newborn Screening." *The Hastings Center Report*, May/June, 38, no. 3 (2008): 24–25. *See* Ruhl, Catherine, and Barbara Moran. "The clinical content of preconception care: preconception care for special populations." *American Journal of Obstetrics and Gynecology* 199, no. 6 (2008): 345–48.
190. Pols, Jeannette. "Towards an Empirical Ethics in Care: Relations with Technologies in Health Care." *Medicine, Health Care and Philosophy Med Health Care and Philos*, (2014), 81–90.
191. Schwab, Abraham Paul. *The Human Microbiome: Ethical, Legal and Social Concerns.* Edited by Rosamond Rhodes and Nada Gligorov. (New York, New York: Oxford University Press, 2013), 38–42.
192. Baily, Mary Ann, and Thomas H. Murray. "Ethics, Evidence, and Cost in Newborn Screening." *The Hastings Center Report*, May/June, 38, no. 3 (2008): 25–26.
193. Hofmann, Bjørn. "The Myth of Technology in Health Care," *Science and Engineering Ethics* 8, no. 1, (2000): 17–28.

194. Baily, Mary Ann, and Thomas H. Murray. "Ethics, Evidence, and Cost in Newborn Screening." *The Hastings Center Report*, May/June, 38, no. 3 (2008): 25–27.

195. Schwab, Abraham Paul. *The Human Microbiome: Ethical, Legal and Social Concerns.* Edited by Rosamond Rhodes and Nada Gligorov. (New York, New York: Oxford University Press, 2013), 23 – 28.

196. Baily, Mary Ann, and Thomas H. Murray. "Ethics, Evidence, and Cost in Newborn Screening." *The Hastings Center Report*, May/June, 38, no. 3 (2008): 27–31.

197. Williams, Janet K., Heather Skirton, and Agnes Masny. "Ethics, Policy, and Educational Issues in Genetic Testing," *Journal of Nursing Scholarship*, 38, no. 2. (2006): 119–20.

198. Williams, Janet K., Heather Skirton, and Agnes Masny. "Ethics, Policy, and Educational Issues in Genetic Testing," *Journal of Nursing Scholarship*, 38, no. 2. (2006): 120–21.

199. Williams, Janet K., Heather Skirton, and Agnes Masny. "Ethics, Policy, and Educational Issues in Genetic Testing," *Journal of Nursing Scholarship*, 38, no. 2. (2006): 121–22.

200. Rothman, David J. Strangers at the Bedside a History of How Law and Bioethics Transformed Medical Decision-making, (New York, NY: Basic Books, 1991), 35–49.

201. Williams, Janet K., Heather Skirton, and Agnes Masny. "Ethics, Policy, and Educational Issues in Genetic Testing," *Journal of Nursing Scholarship*, 38, no. 2. (2006): 122–24.

202. American Academy of Pediatrics, 2001; ASHG/ACMG Report, 1995; Report of a Working Party of the Clinical Genetics Society [UK], 1994.

203. Williams, Janet K., Heather Skirton, and Agnes Masny. "Ethics, Policy, and Educational Issues in Genetic Testing," *Journal of Nursing Scholarship*, 38, no. 2. (2006): 122–124.

# Chapter 2
# Methods and Standards of Clinical Ethics Consultation

Despite the array of consultation methods implemented and used throughout various ethical issues, this text only addresses a few foundational matters that curriculums should possess [1]. This discussion limits consultation methods in this chapter because not all consultation methods are credible or relevant. The mentioned consultation methods are standard in contemporary practices—the procedures selected for this analysis aid in developing a pragmatic amalgamation of theories.

Currently, interpretations and definitions of clinical ethics attempt to demonstrate morally acceptable practices in clinical medicine. However, contemporary defining factors of clinical ethics are inadequate due to their stagnancy in principlism and the moral relativity that accompanies its identification. For instance, while one group believes they are acting ethically by promoting a patient's autonomy, another group may insist that justice is the principle that supersedes individual independence [2]. Principlism becomes deadlocked in moral discussions because it begs questions of ethical priority. Naturally, various theorists argue that each principle or pillar of bioethics requires dynamic balance [3].In this analysis, we demonstrate the effectiveness of various ethical theories through amalgamation and theoretical extraction. In doing so, this discussion reflects the efficacy of comparative approaches in health care ethics. Though bioethical principles may not serve as the mechanism in which ethics consults occur, they are the foundation of all moral decision-making in health care ethics.

Although definitions of clinical ethics rooted in principlism aim to benefit patients in ethically questionable situations, a far more appropriate and beneficial definition of clinical ethics may be uncovered by amalgamating the ASBH's goals and definition of healthcare ethics consultation with Bernard Lo's description in his text, *Resolving Ethical Dilemmas: A Guide for Clinicians*. Although Lo does not give a concrete definition of clinical ethics, he describes clinical ethics and the intricacies that accompany its practice. According to Lo, clinical ethics differentiates itself from traditional bioethics by honing interactive dichotomies and relationships in clinical care [4]. In clinical settings, patients interact with physicians, nurses, and

J. T. Bertino, *Clinical Ethics for Consultation Practice*,
https://doi.org/10.1007/978-3-030-90182-0_2

other medical staff. This facet of clinical care does not necessarily mean that patients foster a relationship with health care professionals or vice versa. The ethical aspects in clinical settings involving a patient-physician relationship require assessing one's value-judgments, the differences between morally acceptable actions, and decisions about effective or safe treatments [5]. Lo's description of clinical ethics presents insights into the definition of clinical ethics. While various relationships in clinical medicine may pertain to action, the relationship clinical ethics forges result in a dichotomy that answers questions of action and inaction [6].

Lo expands his description of clinical ethics by investigating the differences between morality and ethics [7]. He notes that, although these terms are interchangeable, tremendous differences exist between these concepts [8]. In clinical ethics, distinguishing these terms are of the utmost importance. Suppose a clinical ethicist exercises an ethics facilitation approach. The ethicist must adhere to and respect the involved stakeholder's wishes while simultaneously upholding ethical normativity [9]. Morality entails values and beliefs that have no empirical or tangible evidence for their existence [10]. While morality possesses subjective elements, ethics pertains to a formal area of philosophical reasoning that demands argumentative justifications for its import. Morality pertains to individual values, either rooted in a spiritual nature or otherwise, guiding one's behavior. Ethics refers to the traditional process of identifying the *why* and *how* of value-based questions [11].

According to the ASBH core competencies, healthcare ethics consultation is, first and foremost, a service that aids relevant members involved in a healthcare-related discrepancy [12]. Furthermore, the ASBH stipulates that the disparities that arise in healthcare generally concern value-laden concerns of right and wrong [13]. The responsibilities of an ethics consultant differ from the roles and responsibilities of other healthcare professionals due to the consultant's interest in the ethical permissibility of medical practices [14, 15].Ethical questions that arise in healthcare require a definition that answers questions of moral supportability [16]. This point specifies the uniqueness and inherent differences between bioethics and clinical ethics [17].While bioethics houses various disciplines, the ASBH core competencies stipulate that clinical ethics is a unique specialty that manifests in clinical ethics consultation [18].

To further build upon an effective ethics consultation methodology, consider Case 2.1. Various ethical concepts and circumstances are present in Case 2.1. The foundational principles are all present. Still, the desire to do what is best for the patient lies at the center of the conflict. The patient is capacitated and aware of the risks associated with eating by mouth. The patient's family and friends do not necessarily understand the risks of eating by mouth and thus have no reservation for bringing the patient food. In this respect, an ethicist may facilitate education from care teams to families. Families subsequently obtain information that allows them to make a better-informed choice. The patient's family and other involved stakeholders may assist the patient in eating insofar as they both adhere to his will and understand their role in feeding him. With every attempted bite, the patient risks a life-threatening condition. However, it is essential to consider the underlying values of the patient before the ethicist makes formal recommendations.

An ethics facilitation approach seeks a shared understanding of a patient's values and goals. In most cases, team members and surrogates possess a moral obligation to adhere to the patient's wishes and goals. Still, care teams may not always prioritize the patient's autonomy. Instances wherein a patient lacks the necessary decision-making capacity to make an informed choice may result in a deviation from a stated wish. Still, in Case 2.1, the patient possesses the necessary decision-making capacity to make a choice surrounding their clinical circumstance. Asperation presents a risk, but the patient and his family agree that the risk of aspiration is an acceptable burden. Still, knowing the likely outcome of a patient's behavior does not insight confidence regarding the patient's overall safety and preservation. Often, a patient's deviation from medical advice does not sit well with many physicians. After all, hospital settings house professionals that possess far superior clinical knowledge than most patients. A severe deviation from medical advice begs the question of why the patient chose to do so. Considering the benefits and burdens of treatment, the details surrounding quality of life. The patient's decision-making capacity must also receive attention. Still, provided that patients meet these criteria, a principlist standpoint dictates that the patient's wishes be honored.

Comparing and subsequently combining theories proves to be a successful and effective means of uncovering a formal definition of clinical ethics. With Lo and the ASBH's critiques and definitions in mind, a standard definition of clinical ethics begins to form. Clinical ethics is a vocational service wherein moral experts guide patients, families, surrogates, and healthcare professionals through value-laden discrepancies [19]. This definition delineates the role of health care ethics comprehensively. It demonstrates the effectiveness of combining and filtering through multiple considerations [20]. With the model and procedure of comparative amalgamation explained, the analysis now tends to various ethical consultation theories. Applying these theories to a clinical case may elucidate both the theoretical and practical aspects of ethics consultation methods.

Although Case 2.1 does not comprehensively represent clinical ethics issues, it still provides a medium for assessing different models and methods of clinical ethics consultation.

## Methods of Ethics Consultation

The following section investigates three methods of clinical ethics consultation: Process and Format, Four Topics, and CASES. These methods are helpful in contemporary clinical ethics consultation. However, the field has not accepted a universal practice into clinical ethics consultations for various reasons. In many respects, it may not be appropriate to pin down a single ethics consultation methodology. Style, background, and clinical scenarios all affect the methodology in which an ethicist may approach a given situation [21]. Regardless of a chosen consultation method, the most pressing issue in clinical health concerns the various ethical, moral, and practical factors accompanying a clinical case. Furthermore, the methods and practices involved in clinical ethics consultation perform their duties well enough for a consultant to choose whatever manner they see fit. Here, we uncover the details of each consultation method but leave the operational and stylistic approach to the ethicist within their institution.

Process and Format, established by Robert Orr and Wayne Shelton, places a tremendous emphasis on patient documentation and hands-on communication between ethicists and other involved party members [22]. Orr and Shelton emphasize the need for clinical ethics consultation when value-laden issues create discrepancies among involved parties [23]. Process and Format is beneficial when addressing contentious stakeholder dynamics [24]. However, certain cases may not require a tremendous emphasis on patient involvement. Discrepancies often occur when patient and stakeholder values conflict or when a patient's values are unknown. These instances benefit from ethical facilitation [25]. These instances include cases where the patient is a newborn or perhaps cognitively compromised [26]. This issue is problematic because the Process and Format approach does not necessarily address problems of substitutive judgment, nor does Process and Format regulate options for situations of this magnitude [27]. Despite this shortcoming, Orr and Shelton give detailed and helpful information regarding patient interactions and family members [28]. Process and Format dictates that it is almost always appropriate to visit a patient involved in an ethics consultation. Still, there are instances where patient involvement is not warranted [29]. Prominent examples include clinical scenarios wherein the patient is obtunded or incapable of engaging in clinical discussions.

Process and Format dictates that the ethics consultant properly introduce themselves and clearly state their role in the hospital [30].To uphold this aspect of Process and Format, instances where the patient lacks decision-making capacity require the consultant to speak with family members and ask questions about the patient, *i.e.*, their personality, what they like to do, specific hobbies, and other facets of the patient's life [31]. These questions present the family with a sense of familiarity and aid in ethics facilitation by understanding the patient, including insight into the patient's baseline status [32]. In Case 2.1, there is a clear divide between the patient's lifestyle and his illness's limitations on his wishes. The patient enjoyed eating certain kinds of foods, a tremendous cooking enthusiast and lover of world cuisine. Naturally, this aspect of his life was a facet he did not want to compromise, despite his illness [33].

On the one hand, the patient could continue his course of medical treatment, adhering to the physician's orders to take no nutrition by mouth. Adhering to the physician's medical advice accomplishes a foundational goal in medicine. On the other hand, the patient could choose to pursue comfort options and receive at-home Hospice care until an aspiration event or other event takes the patient's life. Since this option is not a curative one, the patient's care may alter in a way that allows him to live a lifestyle of his choosing [34].

By implementing the Process and Format approach to Case 2.1, various qualitative details about the patient's wishes have a venue in which they receive consideration. Process and Format is especially relevant for cases involving patients who are apprehensive about accepting comfort measures. Naturally, the decision to cease aggressive treatments and pursue comfort measures is complicated. After all, patients who enroll in services like hospice are individuals who inevitably must come to terms with the last stages of life. However, Process and Format proves helpful by asserting that this consultation method does not need to present itself as a method that imposes treatment or forces beneficence. Instead, Process and Format uncovers multiple options for patients in an autonomous fashion. Rather than granting decisional priority to a health care professional, this consultation module allows patients to review qualitative options and decide accordingly. However, in cases like 2.1, autonomous choices become difficult when determining comfort measures. Progress in autonomous decision-making requires a patient's ability to consent to treatment and a health care professional's ability to relay information.

Another beneficial aspect of Process and Format involves giving detailed and valuable information regarding interactions between patients and family members [35]. Process and Format states that it is almost always appropriate to visit patients, with certain exceptions [36]. However, in Case 2.1, direct communication is precious. Fortunately, the patient possesses the capacity to consent to treatment.

Process and Format insist that the ethics consultant properly introduce themselves and clearly state their role [37]. Furthermore, in cases like 2.1, the consultant should speak with the patient and family members and ask questions about the inherent values that underlie the shared interests of stakeholders. What are the patient's interests? What is important to the patient? [38] In Case 2.1, it is clear that the patient's passion for food directly influenced his decision to move to a palliative plan. These questions give the family a sense of familiarity and aid in ethics facilitation by understanding the patient's wishes [39].

The Four Topics method is an incredibly beneficial method for addressing stakeholder value-laden discrepancies. This methodology heavily depends on principlism and reflects an accurate summation of how medical professionals approach patient care. Still, patients and their surrogates are often unfamiliar with medical practice and clinical ethics and leave sparse argumentative evidence for their claims [40].

## Four Quadrants

Case 2.1 presents a conundrum in clinical ethics—specifically, a conflict between patients' autonomous decision-making and medical beneficence. On the one hand, the care team could allow the patient to exercise his autonomy by not imposing treatments like mechanical feeding. Medical practice procedures develop around patient safety. Preventing harm and aiming toward a good end for an individual is at the heart of medical care. The difficulty surrounding this option involves usurping The patient's autonomy in the name of beneficence [41].

A critical theme within Case 2.1 involves the informative and educational obligations of clinicians [42]. Physicians must prepare to dictate information in an educational and informative manner [43, 44]. A care team's obligation to protect, heal and respect patients will inevitably conflict with others involved, including the patient.

1. Medical Indications: This quadrant bases itself on beneficence and non-maleficence [45]. Typically, medical professionals uncover the medical indications regarding a patient's medical problem, treatment goals, and the risks and benefits of the patient's course of treatment [46].Regardless of a patient's available treatment options—curative or otherwise—the aim of treatment requires clear delineation of the possible [47]. In doing so, determinations about the risks and benefits of a procedure become elaborated. For instance, if a terminally ill patient's goal is to extend their life as long as possible, strictly adhering to comfort options or options involving terminal weaning is not appropriate. Alternatively, if the patient's goals include enjoying their remaining years without the difficulties of low-yield high-burden treatments, comfort options and palliative efforts become appropriate [48].

   Although stated subtly, the second quadrant of Four Topics notes two standards for surrogate decision-making: substitutive judgment and best interest [49]. However, unlike Process and Format, Four Topics deem surrogate decision-making a formal indication by a consultant rather than a constructive conversation between a consultant and a family member [50]. Both consultation methods contain beneficial aspects in their application. While allocating patient decision-making to surrogates and obtaining pertinent information regarding a patient's wishes are well represented in both approaches, it is still unclear how educational and informative aspects regarding treatment manifest. The next quadrant of Four Topics is especially pertinent to The patient's case and provides insights into comfort measures' informative elements.

2. Preferences of Patients: The second quadrant of Four Topics, patients' preferences, is rooted in the principle of autonomy for apparent reasons. Patients' preferences and beliefs are inherent indications of autonomy and exercise personal autonomy through the medium of choice [51]. The second quadrant first addresses whether the patient has received information about their diagnosis treatment options and the benefits and risks accompanying these options [52]. Regarding clinical ethics consultations surrounding terminally ill patients, the second quadrant of Four Topics may apply to the relevant information concerning the possible outcomes and expectations of curative or comfort options. However, the mental capacity of patients who face decisions at the end of life may not have their preferences clearly articulated. Understanding a patient's preferences before incapacitation is a crucial step in the advance care planning process. An appropriate surrogate appointment is also critical if the team seeks to honor the patient's wishes to the best of their ability [53]. Identifying a surrogate decision-maker is a relatively simple task, yet Four Topics only asks what standards govern the surrogate's decisions and does not facilitate discussions with a disgruntled or unreasonable surrogate decision-maker [54].

Suppose the patient in Case 2.1 is obtunded, and their surrogate asks for interventions that are considered harmful or not medically indicated. While the surrogate wants what is best for the patient, unreasonable requests are often viewed as futile interventions and commonly perpetuate misconceptions about decision-making capacity [55].In contrast, it is impossible to determine the patient's preferences with certainty—items like advance directives aid this process. Unfortunately, while most Americans do not possess an advance directive, the decision-making process is left to surrogates determined by state laws surrounding the next of kin statutes. Since this process is onerous, ethical involvement through Four Topics aids the overall conversation and facilitates dialogue effectively.

Although stated subtly, the second quadrant of Four Topics notes two standards for surrogate decision-making: substitutive judgment and best interest. Unlike Process and Format, Four Topics deem surrogate decision-making a formal indication by consultation rather than a constructive conversation between a consultant and a family member. In this respect, a hybrid combination of both Process and Format and Four Quadrants would be highly beneficial in clinical ethics cases that involve surrogate decision-making [56].In a commentary on beneficence, autonomy, and their relationship with futility, Dr. Kenneth Prager notes that treatment goals for a case like 2.1 require clear delineation [57]. In doing so, determinations about harm, healing, and futility are determined. For instance, if the patient's goal is to delay his death as long as possible, his original course of treatment is appropriate. However, since the patient reevaluated his goals and determined that he wanted to live the rest of his life comfortably, the current treatment was inappropriate and even considered futile [58]. In this respect, the first quadrant of Four Topics aids the ethical assessment of The patient by actively seeking and reevaluating his treatment goals [59].

3. Quality of Life: The third quadrant of Four Topics involves the quality of life inquiries. Gaining insight into a patient's preferred course of treatment is primarily attributed to their perceived quality of life. Obtaining sufficient informed consent from patients is all the more important insofar as patients must determine an acceptable life [60].The third quadrant relies on the principle of beneficence and non-maleficence [61]. Regarding The patient's case, it is clear that quality of life is a reasonable issue that may aid in determining comfort decisions, withdrawing treatment, or implement other palliative interventions. This facet of Four Topics addresses the possible biases and interpretations of life quality but does not specify who may have these biases and how they become addressed [62].

Though the patient has made their wish clear, loved ones and care providers alike may feel this autonomous choice is irrational or inappropriate [63]. However, the involved stakeholders are not in the same situation as the patient, nor have they likely experienced the patient's clinical situation firsthand. In this respect, surrogate decision-making requires significant empathy and insight into a patient's condition.

4. Contextual Features: The final quadrant entails logistical and professional standards of practice [64]. Regarding Case 2.1, the professional standards that a
physician must uphold are incredibly relevant. However, the physician's
approach and his actions' permissibility depend on this case's methodological
approach. While a physician must protect a patient from harm, the physician
must still uphold a patient's autonomy. However, in Case 2.1, the patient's goals
do not necessarily align with the treatment plan [65]. The team establishes that
the patient has a high aspiration risk and will likely die if he encounters another
episode of aspiration pneumonia, yet he still insists on eating by mouth.
Contextual features aid in identifying the situational parameters of value discrepancies in clinical cases.

In bioethical instances, care providers broadly understand that a patient's
declared wishes should be honored as best as possible, not honored at all costs.
Through a Four Topics Approach, any health care provider can assess a patient-
based clinical ethics consult. Each quadrant is taking the foundational principles of
bioethics and applying a vocational approach. A vocational approach to ethics benefits health care providers but only scratches the surface of moral understanding
clinical ethics professionals should know.

In recent years, informed consent has become a standard of ethical practice in
medicine due to its effectiveness in preventing patient autonomy violations.
Informed consent is an operational procedure that ensures a patient assents to an
individual or a series of medical interventions. In this process, the acting physician
possesses a moral and professional responsibility to discuss all relevant information
regarding the medical procedure with the patient [66]. Furthermore, proper execu-
tion of informed consent establishes an agreement that stipulates penalties for physicians or medical professionals that deviate from the plans and interventions
discussed with a patient. While the final decisions are somewhat authoritative, a
patient's involvement in approving a physician's decisions stems from the realm of
respect for personhood and uphold communication between the healer and the
patient [67].

Since informed consent is the act of respecting a patient's autonomy by providing relevant information regarding medical treatment, articulating relevant information about a diagnosis and treatment plan is paramount [68]. In Case 2.1, the medical
staff articulate risks and adverse outcomes of eating by mouth. However, the valid
preferences of a patient reside within the patient. Preferences of patients that do not
align with medical advice are noted but often doubted. This phenomenon is mainly
due to the medical team's inherent bias surrounding their professional opinion.
Naturally, medical practitioners endorse and stand by their professional recommendations [69].

It could very well be that the patient's initial plan had no feasible benefits and
even a futile course of action. Communication between the physician and the patient
regarding the patient's diagnosis and treatment plan begins with a physician's diagnostic explanation. However, by discussing the issues surrounding the patient's
inability to swallow safely and the subsequent aspiration pneumonia, the first step

of obtaining informed consent, namely, properly informing a patient, was adequately presented to the patient and his family. Obtaining informed consent is a delicate facet of clinical conversation for various reasons. First, the physician must not use conflated medical jargon or attempt to confuse the patient in any way. The physician's responsibility is to educate the patient on these terms and their relevance to their diagnosis [70]. This detail also aids in avoiding coercive language or statements [71]. Second, the physician must articulate the relevant information to ensure that the report's details develop into sound recommendations coupled with good reasons for their proposal [72].

While the physician is the foremost authority regarding medical facts of a patient's illness, they must articulate the reasons for his diagnosis and treatment procedures [73]. At the very least, this step in obtaining informed consent aids in establishing a trustworthy relationship between the physician and the patient. Although the patient is often not versed in medical science, the patient is still the sole decision-maker regarding interventions [74]. Finally, proper communication between the physician and the patient yields a mutually beneficial agreement between both parties [75]. By fully informing a patient and obtaining the patient's consent, the physician may use his diagnostic knowledge appropriately and continuously [76]. However, if the physicians coerced the patient into choosing an appropriate action and prevented harm, coincided with the physician's beliefs rather than The patient's, then the course of treatment is not ethically supportable. Informed consent at its root is a request for acceptance. Once a physician adequately discloses relevant information to a patient, the patient may accept or reject recommendations.

Furthermore, the recommendations presented by a physician must also include a weighing of risks and benefits [77]. This disclosure aspect is essential when obtaining informed consent because specific procedures' risks and benefits may directly influence a patient's permission. While diagnostic and procedural information presented to patients may be accurate, the patient's acceptance is only warranted if they understand the physician's recommendations and the risks and benefits of the proposals [78].To assess the proper disclosure standards, a physician may determine what information is appropriate to disclose based on the "reasonable patient standard." [79] In other words, health care professionals must make informed judgments regarding the manner and extent of information disclosure. If the patient is seemingly reasonable and competent, the physician may deliver all necessary information.

The 'reasonable patient standard' is one of the most common standards regarding informed consent [80]. However, a standard of disclosure that is increasing in popularity is the "subjective" standard [81]. This standard is adequate because it can work on a patient-to-patient level by assessing individual cases with tailored information based on a patient's needs [82]. The subjective norm is enticing from an ethical perspective because this standard facilitates mutual respect and understanding. This latter approach to assessing a patient's standard of competence is especially relevant to The patient's case due to the subjective nature of beneficence that the patient's medical team placed upon him. Although many critics may claim they care team possesses a moral obligation to prevent the patient from eating, the function of

American-based clinical ethics primarily focuses on patient autonomy. It is essential to note that the team's decision to adhere to the patient's wishes over non-maleficence is in itself a beneficent act. To demonstrate this point, a brief consideration of the theory of situation ethics aids in elucidating alternative methods to principlism when determining end-of-life care [83].

Despite ethical advancements in clinical practice, issues remain unresolved—specifically, the limitations of adhering strictly to a principlistic structure. For instance, some critics may argue that the care team placed too much interest in the wrong principle in Case 2.1. Alternatively, Case 2.1 demonstrates ethical fortitude on behalf of the care team due to their emphasis on patient autonomy [84].To promote patient autonomy and ethical practice in medicine, infrastructures like informed consent exist to ensure safe and informed practices, but no unifying rule resolves ethical issues in clinical care [85]. Rather than attempt to develop a practice that can unify divided standards of ethical responsibilities, there may be an instantiation of a set of rules and formulations that aid in moral theory [86].

Under Kantian deontological thinking, various ethical maxims or rules aid in ethical decision-making [87].[1]In this respect, obligation and duty are tantamount. However, deontological maxims lie in their rigidity and stagnancy when dealing with multi-faceted ethical situations. In this respect, ethical principlism becomes a problematic facet of consultation for two reasons. First, ethical principlism mutually excludes facets of itself when attempting to arrive at ethical resolutions [88]. This issue manifests in Case 2.1 with a debate between upholding patient autonomy and maintaining the principle of beneficence and non-maleficence. Second, the rigidity of the four principles restricts practical ethics consultation methods by forcing consultation methods to abide by principles. A possible remedy for accommodating ethical variance is situation ethics. Established by moral theologian Joseph Fletcher, Situation ethics is an ethical approach that understands how ethical situations manifest differently [89]. Although the vast differences between human beings and their clinical situations serve as a primary example of autonomy, an ethical approach that can adapt to their differences requires investigation. Situation ethics does not limit its practice to a singularity and does not apply standards and ethical norms to all ethical instances [90]. Situation ethics is especially keen to human needs and varying situations because this method responds out of affection and empathy [91]. Situation ethics bases its practice upon love and care for individuals and their plights.

Regarding Case 2.1, one can argue that the care team's actions and responses to the patient's behavior come from a place of concern and safety. Though at times frustrating, patients who do not adhere to medical advice or choose a care plan that does not align with their goals can spark doubt in care providers [92]. In addition to this point, perhaps the most controversial philosophical claim from a situation ethics perspective is that love is always and everywhere a morally and ethically good

---

[1] Kant's moral theory and other noramative moral philosophies receive specific attention in chapter three.

determination [93]. This claim is controversial due to the boldness entailed within this claim. However, acting out of love in all situations reduces argumentative conundrums that govern medical complexity, provided this claim is accurate [94]. While principlism combats itself by determining which principle prioritizes ethical decision-making, situation ethics decides the most ethical option based on the decision that yields the most care [95]. If viewed from a situationist ethics standpoint, adhering to the patient's wish is also ethically supportable. By effectively implementing a situation ethics approach, the physicians' conversations with the patient and his family in Case 2.1 may turn toward a standard of care that justifies the patient's desire to eat food by mouth and develop care strategies that do not conflict with the patient's overall goals [96]. Situation ethics is a more effective foundation for clinical ethics consultation models due to its malleability [97]. If appropriately exercised, situation ethics effectively bolster current and future models of ethics consultation.

With the foundational aspects of situation ethics elaborated, the analysis examines a final ethics consultation method. Both Process and Format and Four Topics provide excellent methods for accruing ethically relevant information when constructing a clinical ethics consultation. However, the discussion has also demonstrated the issues accompanying these methods when implemented from a principlism model. Ultimately, the patient in Case 2.1 requires a method that articulates ethical discrepancies within a clinical context and aids in perpetuating a practical attitude toward involved party members. Furthermore, an appropriate consultation method for Case 2.1 must also entail a module that perpetuates appropriate educational aspects for patients and their families.

## CASES

The CASES approach is a valuable model within the clinical context of Case 2.1 due to its ability to adapt a situation ethics framework. First developed by the National Center for Ethics in Health Care, CASES presents a formal checklist criterion for clinical ethics consultations, a property represented well in Process and Format, and Four Topics. Unlike Process and Format and Four Topics, CASES presents a general attitude that other clinical consultation methods can adopt [98].

Clarify, assemble, synthesize, explain, and support are the five elements of the CASES approach to clinical ethics consultation. The first step, clarify, involves uncovering the nature and category of the ethics consultation request, gaining preliminary information about a case, and provides an opportunity to determine what ethics questions are present but not yet recognized or identified by the stakeholders involved [99]. This aspect of CASES is synonymous with the medical indication quadrant of Four Topics but differs in that an assessment of a proper consultation method is adopted. CASES is open-ended enough to apply to various ethical issues and increases the overall clinical ethics consultation goals [100].

In most clinical instances, patients and surrogates may not understand the clinical circumstances surrounding their stay in the hospital, let alone the ethics issues. Here, the ethicist may establish a baseline assessment of the value-laden elements of a case. The clarification aspect of CASES justifies the patient's decision to pursue a specific lifestyle. Objectively, eating by mouth in this clinical instance is not safe, but safety at the end of life or with an end-stage chronic illness often conflicts with what a patient may want. These clinical instances are exceptions to any semblance of rules in clinical ethics. Put bluntly. We only die once. The clinical and social context surround and end of life situation for any human individual should receive specific attention to the patient's preferences [101].

Though this analysis intermittently scrutinizes adhering to a strict principlistic approach to clinical ethics, it is simply unavoidable not to recognize that each bioethical pillar is an individual constant throughout health care. Reasonably, few individuals may argue with the importance of each pillar, though the balance and mechanisms of prioritizing an individual pillar plague the field.

The second and third aspects of CASES, assemble and synthesize, aid in gathering all necessary information regarding a case and determining whether a formal ethics meeting is required [102]. These steps include acquiring the types of information needed, sources of information, a summary of the ethics questions, and identifying an appropriate decision-maker [103]. These steps may appropriately summarize the entirety of what Process and Format and Four Topics accomplish [104]. In Case 2.1, an ethics consultation is undoubtedly justified. Ethics consultants who understand these concepts also need to implement these methodologies in their consults. Consultants who understand consultation methodology may not know how to implement methodologies into their practice. Running through a list of value considerations and discrepancies is an objective mechanism that may promote thinking, but ethics consultation requires practical application in an organic manner. The methodologies mentioned in this analysis are necessary for an ethicist. The communicative skills required to use effective consultation methodology require practice and controlled assessment. Ethics consultation simulation is an effective way of accomplishing this training process. Though individual hospital systems may use staff to simulate or reenact a clinical ethics case, bringing outside actors can provide a sense of authenticity to the simulation [105].

The fourth and fifth steps of CASES include explanation and support [106]. These two facets set the CASES methodology apart from others by insisting on immediate steps that require the most priority. Within the context of explanation, consultants must communicate the synthesized aspect of the consultation with key participants [107].Communication with stakeholders in clinical ethics consultation requires a level of stylistic fluidity. Still, the case methodologies presented in this analysis are templates for a greater purpose [108].

The consultation methodologies mentioned in this section are roadmaps for critical thinking. They are not absolutes. Though this analysis argues for a standard ethics consultation curriculum, it does not discourage critical thinking. Ethics consultation methodologies provide structure and meaning to one's thinking, but they cannot stand alone. The malleability of these methodologies is equally as crucial as the ethicist's ability to adapt, too. In subsequent sections, we examine the significant current efforts on professional ethics consultation accreditation [109].

## ASBH and CECA Efforts Toward Accreditation

In 2009, the ASBH formed the Clinical Ethics Consultation Affairs (CECA) standing committee. This committee was formed to address individuals' competency and professional knowledge responsible for providing clinical ethics consultations to patients, families, and health care professionals. These concerns primarily involve the legitimacy of consultations and consultant competence. Furthermore, this committee was formed to improve both basic and advanced levels of competency for clinical ethics consultants based on the ASBH's *Competencies for Health Care Ethics Consultation* [110]. The evaluation is broken into two parts. Part I presents the ASBH's pilot program and the CECA subcommittee's recommendations to the directors' board. Part II discusses appendix B of the CECA report. Appendix B outlines specific skill and knowledge areas that must be met for an individual to receive a sufficient clinical ethics consultation education and accreditation. Appendix C gives examples of certifying bodies currently using some facets of the ASBH's pilot program. Although this information is helpful in that accrediting bodies to demonstrate the topics listed in this report, this critique examines the details of both the ASBH's pilot program for professional ethics accreditation and the CECA's recommendations for the ASBH's pilot program via the CECA's 2010 report to the ASBH board of directors. In conjunction with one another, this critique demonstrates current methods of evaluating the levels of competence clinical health care ethicists possess by evaluating the ASBH's pilot program [111].

The report begins by explaining the inherent demand for qualified clinical ethics consultants. The primary question the report addresses is whether individuals conducting clinical ethics consultation possess adequate qualifications. According to Fox and Colleagues' national survey, the report mentions that only 5% of consultants have completed a fellowship or graduate program in bioethics [112]. Despite this statistic, the ASBH and CECA have emphasized other means of accreditation and certification for clinical ethics consultants due to graduate programs' legitimacy. According to the ASBH and CECA, no bioethics graduate or fellowship program possesses a standard of education that is agreed upon and recognized as a universal accreditation standard. In this respect, the ASBH asserts that no formal assessment and attestation of an individual's clinical ethics consultation competencies exists [113].Initially, the ASBH developed a multiple-choice examination to determine the competency of clinical ethics consultants and other individuals assisting with ethics consultations. This exam comprises multiple-choice questions that test a range of topics involved with HCEC, including bedside manner, role delineations, and necessary skill sets established by the ASBH's *Core Competencies for*

*Healthcare Ethics Consultation* and *Improving Competencies in Clinical Ethics Consultation.* However, the CECA committee indicates that the multiple-choice examination that the ASBH requires is an insufficient means of testing competencies and skills for clinical ethics consultants [114].

In conjunction with the ASBH's requirements, the CECA suggests further actions involved with an examination process. These actions include a multiple-choice examination including an essay, a written case study analysis that involves an example of electronic medical record (EMR) documentation, and an oral interview with, presumably, an experienced and skilled consultant. Additionally, the CECA advocates for five additional methods to measure CEC skills and knowledge competencies. These methods include an evaluation based on mock consultations, evidence of having performed a minimum number of consultations as a lead consultant, a graduate degree in the applicant's field, formal evidence of clinical ethics consultation education and training, *i.e.*, a bioethics degree program, certification program, or continuing education, and a letter of recommendation from a supervisor or colleague who has provided clinical ethics consultation experience and observed the applicant's consultation skills [115].

Following the modifications mentioned above to the ASBH's examination process, the CECA subcommittee outlines five necessary components for an individual to receive a standardized accreditation for conducting clinical ethics consultations. The first certification point involves a written exam. This exam has been discussed in the previous section and involves a much more detailed approach to certification than the previous standing multiple-choice examination composed by the ASBH. This examination involves a multiple-choice examination that tests an individual's essential competencies in clinical ethics and clinical ethics consultations and requires individuals to demonstrate their writing skills via written essay sections that include a written case study analysis. Furthermore, the exam also tests the written fortitude of applicants via EMR evaluations [116]. The second certification point involves a portfolio that applicants must provide. This portfolio must include summaries of a minimum number of ethics consultations that have been conducted in the past year, *i.e.*, anonymous EMR documentation of three case consultations in the prior fiscal year [117]. This aspect of the application process assesses the writing capabilities of the applicant and their ability to curtail their writing specifically for clinical ethics consultation. The third certification point the CECA presents involves an observational element to the certification and application. While this point is undeveloped in the CECA and ASBH's report, this aspect of the application process encourages eye-witness testimony to the effectiveness of an individual's clinical ethics consultation skills. Although it is not explicitly outlined in the report, the individual performing the evaluation is presumably an experienced consultant who possesses the skills to perform consultations and evaluate prospective consultants. Naturally, this provokes further operational questions surrounding one's ability to critique other ethicists [118].The fourth certification point involves letters of reference with attention to "360-degree reviews" from supervisors, colleagues, and other affiliated stakeholders. A supervisor, for instance, may serve as a director of a clinical ethics consultation service or an ethics committee chair. At the same time, a collogue is an individual who has personally observed the applicant during a consultation. Two other individuals who conduct a "360-degree review" may entail a subordinate who has observed an applicant provide consultation or even a patient or family member previously involved with an ethics consultation. If available, the latter individual may present a standardized form that evaluates the performance of a specific consultation with a narrative explanation [119]. Finally, the fifth point involves an interview by a panel of experienced clinical ethics consultants [120].

The CECA then presents a list of five recommendations. These recommendations attempt to reconcile discrepancies and possible issues from the ASBH's application and accreditation program for consultants. The CECA mentions a concern in their recommendations to the board of directors of the ASBH regarding companies that provide test development and implementation. Although these companies are not unfamiliar with start-up testing and implementation methods, maintenance costs and start-up overhead costs can be exponential and thus require attention. The money used to develop the necessary examinations for certification and accreditation must be composed fairly while simultaneously covering the expansive knowledge base involved in clinical ethics consultation. Furthermore, these companies are responsible for composing universal templates for the portfolio mentioned above, reference letters, and observational standards [121].

The CECA notes that national organizations spend over 100,000 dollars annually to maintain certification testing and compliance standards from companies that typically monitor and produce standardized testing. In comparison, organizations spend over 500,000 dollars in testing research and development alone. The CECA recommends that the ASBH pursue funding to cover start-up costs for testing development to reconcile this cost discrepancy. Since the certification process that the ASBH and CECA have developed requires rigorous attention and detail, there can only be minimal restrictions concerning the cost of research and development. Although many individuals believe there is not a high enough demand to support or justify the start-up costs of a testing program, the CECA and ASBH believe that demand for standardized testing and certification will increase once the community reaches consensus on a formal process [122].

The CECA concludes this point by posing five questions. These questions investigate key issues surrounding start-up costs for external test development companies. These questions include:

1. How many individuals providing clinical ethics consultation are likely willing to receive certification themselves?
2. What cost differences manifest for individuals who must pay for self-funded testing?
3. What are the pros and cons of outsourcing certification versus internal certification through the ASBH?
4. What are the liability implications for clinical ethics consultants who do or do not receive certification?
5. How should clinical ethics certification aid and work in conjunction with graduate programs?

All five of the questions mentioned above require answers before the implementation of a formal testing process. Furthermore, the testing process proposed by the ASBH and modified by the CECA subcommittee cannot come to fruition without formidable answers to the above questions. While question two pertains to the financial concerns of the certification project, questions one, three, four, and five all focus on the practicality of the pilot program [123].The CECA sub-committee determined that the ASBH must certify individuals who provide clinical ethics consultation services at various competency levels. Individuals providing clinical ethics consultation across hospital systems and other venues do not align in their understanding of clinical ethics or their education methods. The CECA notes that providing clinical ethics consultation must receive thorough and advanced competency training that comprehensively sifts through information using the previous evaluation methods. The CECA deem their evaluation methods as methodically rigorous processes that can open a practical pathway to a self-learning program that teaches and demonstrates basic clinical ethics consultation knowledge for individuals providing consultations in a team model. In doing so, a wide range of individuals may receive training in a short amount of time due to the practicality of transferable information through mediums like online classes. These programs encourage ethics consultants to self-educate themselves with tools provided by the ASBH [124]. Presumably, the tools provided by the ASBH are adequate and ensure advanced educational points for those looking to continue their work as clinical ethics consultants. The CECA and ASBH wish to relay specific knowledge points that can accommodate online and team-oriented self-learning programs. Most of the information provided to students who seek online support and education derives from the ASBH's *Education Guide,* which provides basic clinical ethics consultation knowledge for individuals providing consultations at a basic level [125].

Assuming individuals complete this course, students still cannot achieve equivalency in becoming certified clinical ethics consultants. The online self-learning programs are synonymous with the certified IRB professional exam demonstrating the advanced expertise of research ethics. In contrast, completion of various modules demonstrates basic knowledge of research ethics. However, the CECA proposes a fundamental difference in certification and education for these individuals in that the clinical ethics certification process evaluates more than elementary aptitude [126]. The CECA concludes this section by advocating for the ASBH's exploration into licensing primary educational products to generate revenue to help fund the certification process. Thus, proper funding acquisition is a necessary component for the ASBH's vision to come to fruition [127].

Three issues that develop from the CECA's recommendations to the ASBH involve grandparenting individuals who already possess advanced experience and knowledge in clinical ethics consultation, the issues surrounding the establishment of councils that accredit educational programs that the ASBH uses,[2] and the demands

---

[2] Programs that use the ASBH core competencies as an educational foundation for their curriculums.

for certified clinical ethics consultants. By examining the process of grandparenting clinicians and the ASBH's implementation of standards, curriculum developers attain a greater understanding of what clinical ethics programs require. Naturally, the transition period between the inception of the ASBH's new program and the implementation of a new program as a mandated platform requires transition tactics that do not disrupt the work and progress of established ethics consultants who have been practicing clinical ethics for years. Furthermore, these individuals are knowledgeable about the fundamental skills and competencies expected of a professional ethicist. They thus may not need to undergo the same stringent features of the program as other individuals.

The CECA committee believes it is unwise to create a system for grandparenting current clinical ethics consultants. Precisely, the CECA's reservations lie in concerns about premature mandates on these individuals. The CECA notes that early applicants for certification will voluntarily demonstrate their clinical ethics consultation experiences without a mandate to obtain clinical ethics certification [128]. Questions regarding the legitimacy of volunteer certification for advanced practitioners require attention since this presents a risk of wasted time and effort on the practitioner's part if their efforts fail in the face of new mandatory legislation [129]. The final two recommendations presented by the CECA involve considerations surrounding the development of councils for accrediting educational programs that use the ASBH core competencies for teaching, evaluating, and generating demands for certified clinical ethics consultants. Both issues are extremely pressing due to their influence overdeveloping a certification program [130]. Establishing a council that accredits educational programs, albeit graduate or fellowship, must serve as an intermediary step toward accrediting programs that educate and train clinical ethics consultants [131]. The ASBH should consider using the *Core Competencies* as a baseline for clinical ethics knowledge and skill development. This council may also explore continuing education units for education programs that assess knowledge and skill competencies [132].

Finally, the ASBH and CECA's efforts to establish accreditation programs may go wanting if there is no demand for certified clinical ethics consultants. Accrediting bodies, *i.e.,* Joint Commission and other professional organizations, should be aware of the demand for clinical ethics consultants by administering surveys throughout their care systems. These surveys should address patient populations, physicians, and other health professionals. In doing so, there must be a significant demand for more educated consultants that are willing to provide advanced knowledge to families, boards of directors, and patients [133].

The information above outlines the important recommendations the CECA has provided to the ASBH regarding their pilot program for clinical ethics certification and accreditation. With these essential considerations in mind, the report then moves onto detailed strategic plans and information in certification programs, namely, knowledge skill sets for individuals conducting ethics consultations. By examining these methods and skills to evaluate advanced clinical ethics consultation skills and knowledge points for certification, a better understanding of the pilot program and future effort for certification reveals itself. Finally, the recommendations presented by the CECA sub-committee to the ASBH present an array of critical critiques that ultimately provide insights that enhance the ASBH's model for certification and accreditation of individuals conducting clinical ethics consultations. For example, issuing requests for start-up funding, pursuing certification of individuals at the advanced level through self-learning programs, accommodating grandparenting issues, accreditation councils for educational programs, and generating demands for certified individuals conducting ethics consultations are all relevant recommendations that ultimately better the ASBH's pilot program for accreditation and certification.

The CECA's report of the ASBH's pilot program includes three appendices. Appendix A provides names of CECA committee members, while appendices B and C contain fruitful information regarding accreditation methods, skills, and knowledge areas. Appendix B: *Methods to Evaluate Advanced CEC Skills & Knowledge for Certification* dissects clinical ethics certification's knowledge areas into core skills and corresponding assessment methods.

Appendix B of the CECA report examines the methods used to evaluate advanced clinical ethics skills and knowledge points for certification. The appendix has two main sections: Core Skills and Knowledge for Clinical Ethics and Assessment. While the latter section only lists brief mediums through which individuals become assessed or tested, the former section subdivides into three main subsections: Ethical Assessment Skills, Process Skills, and Interpersonal Skills. Each subsection is then further subdivided into 12 total skill areas. Since the amount of information within the CECA report's appendices is overwhelming and complex, this analysis assesses clinical ethics consultants' core skills and knowledge points more than the actual assessment methods. By assessing the skills and knowledge areas presented by the ASBH and CECA report, the pilot program's components become more precise, and the certifying body's aims are in appendix C. The CECA has presented various skill sets and areas to assess an individual's ability to conduct a beneficial and effective clinical ethics consultation.

Furthermore, the CECA's Format for the appropriate skills must come before assessing knowledge areas concerning clinical ethics consultation. The CECA's order of appendices explicitly indicates this. Specifically, the skill areas associated with clinical ethics consultation appear before the knowledge areas. This section deals explicitly with Ethical Assessment Skills, or the first set of skills under evaluation methods [134].The CECA categorizes ethical assessment skills necessary to identify the value or uncertainty of conflict that demands proper ethics consultation. These skills require one's ability to discern and assemble relevant data that is pertinent to a case. The data may range from information gathered from a clinical setting or a less conventional setting like a psychosocial setting. These skills entail assessing a patient's decisional capacity and how this impacts an ethics consultation. The CECA notes that the clinical ethicist documents in the patient's medical record. The consultant must then assess the social and interpersonal dynamics between the patient and other important stakeholders, including family, friends, and medical staff.

Furthermore, distinguishing the ethical dimensions of a case while clearly articulating the ethical and practical concerns are pivotal components that aid in identifying various assumptions brought into a case. Additionally, specific ethical dimensions of a case reveal inherent values embedded within involved stakeholders [135]. Finally, a critical skill that the CECA and ASBH deem necessary for appropriate clinical ethics consultation involves a consultant's ability to identify their relevant moral values and the institution's values. In doing so, the consultant may assess how these values may affect their decision [136].

The skills mentioned above by the CECA and ASBH contain various assessment methods ranging from documentation to written analyses and interviews. All mediums mentioned by the CECA involve some form of testing or reviewable material. The section continues with additional skills that fall under the Ethical Assessment Skill category. Specifically, one of the primary skills a clinical ethics consultant must possess involves the necessity of an individual's ability to analyze the value, uncertainty, or conflict between involved members [137]. For a clinical ethics consultant to demonstrate competence, one must access relevant ethics knowledge and clarify the concepts critically for family members and patients. These concepts include confidentiality concerns, privacy, informed consent, and best interest standards. In doing so, clinical ethics consultants can critically evaluate a situation and subsequently use relevant knowledge of bioethics, bioethics law—barring legal advice—and institutional policies to aid in the facilitation process [138].

The CECA report speculates one's ability to critically evaluate and use relevant knowledge of bioethics another step by presenting further criteria for clinical ethics consultants. Specifically, clinical ethics consultants must utilize relevant moral considerations to aid their analysis, identify and justify morally acceptable opinions, and evaluate evidence and arguments supporting or rejecting certain opinions [139]. Furthermore, the CECA recommends that clinical ethics consultants stay active within their community by remaining up-to-date on peer-reviewed clinical and bioethics journals and books. In doing so, clinical ethics consultants will know how to access and implement the information found within their research. Finally, clinical ethics consultants must also recognize and acknowledge their limitations and possible areas of conflict. In doing so, consultants may avoid or limit instances of moral distress and professional burn-out [140].

The skills under the *Ethical Assessment* category are standard practices that should be known and exercised throughout all clinical ethics consultations. Again, the CECA notes that ethicists may address and determine various assessment methods, including case-based written examinations and interviews. The next section of this critique discusses the CECA's Process Skills and Interpersonal Skills. Synonymous with the Ethical Assessment Skills listed in the previous section, the CECA deem process and interpersonal skills pivotal components for clinical ethics consultants and their professional practice. Although the committee indicates that both categories are vital within advanced skillsets for ethics consultants, both process and interpersonal skills require individual training and attention.

Although unique in their application, process and interpersonal skills resemble the methods and standards this analysis proposes. While the critical aspects of effective clinical ethics consultation lie in applying an individual's ability to understand and reason through ethically precarious situations, the methods that the ASBH's CECA subcommittee presents offer core elements of understanding for clinical ethicists and bolster their abilities as clinicians. Process Skills pertain to a consultant's ability to facilitate formal and informal meetings, build moral consensus, utilize institutional structures, and document consults. Identifying key decision-makers and involved party members is essential in facilitating discussion since both individuals partake in the consultation. However, to conduct these meetings constructively, a clinical ethics consultant must set ground rules for formal meetings, express and stay with families while maintaining a professional role, and establish boundaries for themselves and others [141]. The CECA notes that the most crucial process skill within this category is a consultant's ability to create an atmosphere of trust that respects privacy and confidentiality. In doing so, the consultant develops a forum that allows all party members to feel free to express themselves and their concerns about their loved one or the procedures in question [142].

Another process skill presented by the CECA involves a consultant's ability to build moral consensus among party members. In doing so, consultants help individuals analyze a patient's values alongside their own critically and constructively. Furthermore, this skill allows individuals to identify their underlying biases, assumptions, and prejudices while making decisions for their loved ones [143]. Finally, the CECA notes that a consultant's ability to utilize institutional structures and resources and document consults are skills accompanying a clinical ethics consultation. Utilizing institutional structures allows other consultants and health professionals to review initial observations and findings regarding a case [144].Process skills serve as fundamental tools and elements of a clinical ethicist's practice. These skills involve the general work-related skills that a clinical ethicist should know and use throughout his practice. However, while these skills are mandatory aspects of becoming an effective clinical ethicist, they are useless unless interpersonal skills accompany them. There exist various instances where professionals in any given field are superb at their general tasks and work-related functions.

Nevertheless, the same individuals can lack various qualities that allow them to perform their tasks effectively. For clinical ethicists, interpersonal skills are a necessary aspect of the job. The work involved in clinical ethics requires relay information, critical listening, and mutual understanding with providers, patients, and families. Conducting a consultation in health care may be performed in a fashion that allows the completion of mandatory tasks, such as identifying ethical issues and questions, establishing actions, and subsequently providing recommendations. These tasks are ineffective if the consultant does not critically and personably engage with the involved stakeholders. Interpersonal skills are essential in clinical ethics consultations since the nature of ethics consultations inherently involves other individuals. Interpersonal skills involved with clinical ethics consultation require various abilities on the part of the consultant. The consultant must possess superior listening and communication abilities. Ethicists should use these skills used to promote interest, respect, support, and empathy. A consultant's ability to educate parties involved concerning a case's ethical dimensions is also of the utmost importance due to the fragile relationship between patients and families [145].

During clinical ethics consultations, there is no room for communication break-downs, miscommunications, or quarrels. The consultant's ability to present moral arguments to the party members and the party member's views is vital during the documentation process. The importance of relaying moral arguments is helpful when passing cases over to other consultants or revisiting cases must contain detailed information regarding the viewpoints and beliefs of all party members involved [146]. Finally, interpersonal skills must entail a consultant's ability to pro-mote communication between party members and recognize and attend to various barriers to communication. Since clinical ethics consultants are facilitators above all else, these individuals must possess the skills necessary to relay and interpret infor-mation to those who have not received education on the subject matter, those who misinterpret information, and those who allow emotions to sway their decisions [147].The skills mentioned above are all necessary for clinical ethics consultants, according to the CECA. However, the skill sets above are not practical unless they manifest through a knowledge medium upon which a consultant may develop their ideas and recommendations. Prospective and established ethicists alike should not disregard the facets of the CECA report that address process skills and interpersonal skills. These skills are integral parts of clinical ethics consultations, and programs should foster the skills mentioned above throughout a clinical ethicist's education. Like many issues involved with formalizing curriculum components for ethicists, various debate exists surrounding what should and should not become entailed in a clinical ethicist's education. Synonymous with this project's analysis, the CECA's report advocates for the amalgam of practical skills and knowledge points, albeit represented in a different format.

The following sections assess the CECA report's knowledge areas. These knowl-edge areas address the practical facets of information that trained professionals in ethics should possess to perform adequate clinical ethics consultations. While these facets of a clinical ethicist's education are fundamental and necessary, the analysis clarifies the importance of further expanding their ability to identify virtue in his practice. The CECA report's knowledge area section of Appendix C contains vari-ous information that the CECA recommends as minimum content that a consultant should possess. The content ranges from moral reasoning and ethical theory to information regarding local health care institutions' policies and necessary health law information. The following section discusses knowledge areas about moral rea-soning and ethical theory. Additionally, the analysis discusses fundamental bioethi-cal issues and concepts that frequently appear throughout ethics consultations and practical knowledge areas within health care ethics. These knowledge areas include information regarding health care institutions and other practical information regarding contemporary health care. This critique attempts to group these knowl-edge areas due to their rootedness in theory. Finally, assessing these areas demon-strates the moral reasoning present in contemporary bioethics.

The knowledge area section of Appendix C begins with information regarding moral reasoning and ethical theory, and bioethical issues and concepts in contempo-rary health care. The information includes corresponding assessment methods. These assessment methods include written exams, clinical ethics documentation, interviews, applicant information, and various combinations of each method. Clinical ethics consultants must understand ethical concepts and theories and how they relate to bioethics. These concepts include consequentialist and non-consequentialist approaches, Kantian virtue theory, and theological approaches. The CECA recommends that clinical ethics consultants are proficient with primary texts and theoretical ethics: *i.e.*, Beauchamp and Childress' *Principles of Biomedical Ethics* and the principal/caustic theories the text entails [148]. A clinical ethics consultant must know theories of justice. Specifically, consultants should understand theories of justice in contexts that are especially relevant to resource allocation, fair distribution, and access. Finally, consultants must know their role and authority concerning obligations to provide health care [149].Bioethics issues and concepts in ethics consultations must be understood and critically analyzed by clinical ethics consultants. These issues and concepts include a patient's right to health care, self-determination, refusal of treatment, and confidentiality following a patient's right to privacy.

Consultants should also be aware of "positive" and "negative" rights. Naturally, autonomy and informed consent are pivotal elements of knowledge that clinical ethics consultations must possess [150]. These concepts occur across the field of medical ethics and are relevant concepts that must apply to clinical situations such as patients receiving adequate information, voluntary and involuntary differentiations, competence and decision-making capacity, rationality, and instances of paternalism. Bioethics issues and concepts especially pertain to one's ability to understand and relay the concepts of confidentiality, fiduciary relationships between providers and patients, and the exceptions to uphold patient confidentiality. Other issues concerning a consultant's knowledge areas that pertain to confidentiality and professional relationships include disclosure, deception, and the impact these concepts have on a patient's privacy [151].

Since clinical ethics consultants work with the general population, an ethics consultant must know how to aid and act under challenging patient situations. In doing so, the consultant upholds their duties as a responsible health care professional and demonstrates their valor as an ethicist by assisting and addressing patient compliance [152]. In conjunction with patients who may become difficult to address, a consultant's relationship with other professional staff benefits an entire health care system. Specifically, clinical ethicists' professional relationship with other health care professionals should be a positive one wherein both parties mutually uphold the other's rights and position in the hospital. These rights and duties include, but are not limited to, the parameters of conscientious objection and the duty to care. Clinical ethics consultants must also understand how cultural and religious diversity factor into ethics consultations. This point also includes knowledge about biases based on race, gender, disability, and sexuality [153].Finally, clinical ethics consultants must understand End-of-life decision-making and the complications this area contains, including issues about medical futility, quality of life, euthanasia, physician-assisted suicide, DNR/DNI orders, and withholding nutrition and hydration, must all be present and demonstrated by a consultant through a written exam assessment method and peer evaluation [154].

The ethics consultant must also demonstrate proficient knowledge concerning surrogate decision-making involving minors and incapacitated patients. A clinical ethics consultant must also demonstrate proficiency in all aspects of beginning-of-life care. The knowledge areas of begging-of-life care include reproductive technologies, surrogate parenthood, in vitro fertilization, genetic testing, insurance issues concerning maternity, issues surrounding critically ill newborns, conflicts of interest between involved parties, their relationship to a newborn and mother, sterilization, and abortion. These issues involved with end-of-life care alone create various intersections of issues involved with ethics consultations. For instance, health care professionals expect ethics consultants to understand sterilization and how it pertains to various healthcare institutions' policies and procedures. However, other ethical issues include medical research, therapeutic innovation, experimental treatments, and organ donation [155].

Finally, ethics consultants must understand conflicts of interest involving families and health care organizations in service to the critically and chronically ill. Knowledge concerning insurance, resource allocation, triage, and rationing promotes the consultant's duty toward social responsibilities and obligations to society [156]. Clinical ethics consultants must have ample knowledge about health care systems and how they relate to managed care systems and federal systems, but ethics consultants must also possess ample knowledge concerning clinical contexts. Clinical contexts involve an array of information, including the use of basic terms for human anatomy, diseases, and their prognoses, the history of common illnesses, psychological responses to illness, the processes by which health care professionals diagnose illnesses, awareness of the grieving process, emerging technologies, and a basic understanding of how providers deliver care to an array of individuals through various venues. The venues in which ethics consultants must become familiar include local health care institutions where the consultant works. A consultant's workplace details include the organization's mission statement, services, medical research, medical records, human resources, and spiritual health.

An ethics consultant must also become proficient with their local health care institution's policies [157]. These topics include informed consent, conflicts of interest, conscientious objection, confidentiality and privacy, human experimentation, advanced directives, death by neurological criteria, life-saving treatment in adults, and resolutions of intractable disputes. The practical application of these concepts introduces other topics like impaired providers, error disclosure, medical futility, HIV testing, and disclosure. The above concepts also require a consultant to possess the patient and staff populations' beliefs and perspectives. In addition, these populations require a consultant's beliefs and perspectives that affect the care of racial, ethnic, cultural, and religious groups served by the health care institution [158]. The final section of knowledge areas for clinical consultants includes relevant codes of ethics, professional conduct, guidelines of accrediting organizations, and relevant health law. The codes of conduct from relevant professional and local institutions, a consensus of ethical guidelines, standards of the Joint Commission or equivalent governing body, and patient's bill of rights and responsibilities are all facets of professional conduct that clinical ethics consultants should know [159].

## Concluding Remarks

This chapter examines clinical ethics consultation methods through primary literature and practical application to a clinical case. A formalized definition of clinical
ethics forms through the comparative amalgamation of these two mediums, beginning with an analysis of clinical ethics via Bernard Lo's interpretation of clinical
ethics and the ASBH's definition of ethics. The formalized definition of clinical
ethics subsequently provides a venue for describing Case 2.1. The discussion then
introduced three extremely beneficial clinical consultation methods, Process and
Format, Four Topics, and CASES. We subsequently examine the details of Case 2.1
through the lens of three consultation methods and identifies ethically supportable
recommendations. Finally, the discussion examines Joseph Fletcher's situation ethics approach. This discussion ultimately demonstrates that combining ethics consultation methodologies may aid a consultant rather than principlism methods [160].
While the chapter's discussion of these formidable consultation methods explains
the utility of implementing these methods, no formal accreditation body currently
exists that professionalizes the field of clinical ethics consultation in a manner that
assesses the effectiveness of a consultant's consultation methods. In this respect, the
legitimacy of the consultation mentioned above methods has no standard for their
practice.

The CECA report to the board of directors of the ASBH outlines the ASBH's
pilot program and demonstrates various suggestions and points of improvement.
The report insists on developing a curriculum that consists of both skill areas and
knowledge areas. Upon completing various assessment methods, clinical ethics
consultants develop a knowledge base that aids in conducting professional ethics
consultations. Provided an ethics consultant pass all assessment method procedures,
consultants can receive accreditation and professionally practice clinical ethics consultation under the licensure of the ASBH. However, this chapter's curriculum guide
for accreditation omits key skill points that must become apparent in an accreditation process for clinical ethicists. Uncovering these curriculum points and applying
them to a curriculum aids the process of clinical ethics professionalization.

## References

1. Gill, Robin, *Health Care and Christian Ethics*, (Cambridge: Cambridge University Press
   2008), 3–8.
2. Gillon, R., "Ethics needs principles—four can encompass the rest—and respect for autonomy
   should be 'first among equals'," *Journal of Medical Ethics* 29, no. 30, (2003): 307–312.
3. Gillon, R., "Ethics needs principles—four can encompass the rest—and respect for autonomy
   should be 'first among equals'," *Journal of Medical Ethics* 29, no. 30, (2003): 307–312.
4. Lo, Bernard. *Resolving Ethical Dilemmas: A Guide for Clinicians*. 4th ed. (Philadelphia:
   Wolters Kluwer Health/Lippincott Williams & Wilkins, 2009), 4.
5. Lo, Bernard. *Resolving Ethical Dilemmas: A Guide for Clinicians*. 4th ed. (Philadelphia:
   Wolters Kluwer Health/Lippincott Williams & Wilkins, 2009), 4; *See* DeMarco, Rosanna
   F., Donna Gallagher, Lucy Bradley-Springer, Sande Gracia Jones, and Julie Visk.

"Recommendations and reality: Perceived patient, provider, and policy barriers to implementing routine HIV screening and proposed solutions," *Nursing outlook* 60, no. 2 (2012): 72–80.

6. Lo, Bernard, *Resolving Ethical Dilemmas: A Guide for Clinicians*, 4th ed. (Philadelphia: Wolters Kluwer Health/Lippincott Williams & Wilkins, 2009), 4.

7. Lo, Bernard, *Resolving Ethical Dilemmas: A Guide for Clinicians*, 4th ed (Philadelphia: Wolters Kluwer Health/Lippincott Williams & Wilkins, 2009), 4; cf. Beauchamp, Tom L. "The 'four principles' approach to health care ethics." *Principles of health care ethics* (2007): 3–10.

8. Lo, Bernard, *Resolving Ethical Dilemmas: A Guide for Clinicians*, 4th ed (Philadelphia: Wolters Kluwer Health/Lippincott Williams & Wilkins, 2009), 5.

9. Lo, Bernard, *Resolving Ethical Dilemmas: A Guide for Clinicians*, 4th ed (Philadelphia: Wolters Kluwer Health/Lippincott Williams & Wilkins, 2009), 5.

10. Lo, Bernard, *Resolving Ethical Dilemmas: A Guide for Clinicians*, 4th ed (Philadelphia: Wolters Kluwer Health/Lippincott Williams & Wilkins, 2009), 5–7. cf. Pham, Michel Tuan. "Emotion and rationality: A critical review and interpretation of empirical evidence." *Review of general psychology* 11, no. 2 (2007): 155.

11. Post, Linda Farber., Jeffrey Blustein, and Nancy N. Dubler. *Handbook for Health Care Ethics Committees*, (Baltimore: Johns Hopkins University Press, 2007), 11–22. *See* Quill, Timothy E., Robert Arnold, and Anthony L. Back. "Discussing treatment preferences with patients who want "everything"." *Annals of Internal Medicine* 151, no. 5 (2009): 345–49.

12. *Core Competencies for Healthcare Ethics Consultation.* 2nd ed. Glenview, IL: American Society for Bioethics & Humanities, 2012. 2.

13. Arnold, Robert M., MD, Kenneth A. Berkowitz, MD, Nancy Neveloff Dubler, LLB, Denise Dudzinski, PhD MTS, Ellen Fox, MD, Andrea Frolic, PhD, Jacqueline J. Glover, PhD, Kenneth Kipnis, PhD, Ann Marie Natali, MBA, William A. Nelson, PhD, Mary V. Rorty, PhD, Paul M. Schyve, MD, Joy D. Skeel, MDiv, and Anita J. Tarzian, PhD RN. *Core competencies for healthcare ethics consultation: the report of the American Society for Bioethics and Humanities.* 2nd Ed. (Glenview, IL: ASBH, American Society for Bioethics and Humanities, 2011), 2.

14. Arnold, Robert M., MD, Kenneth A. Berkowitz, MD, Nancy Neveloff Dubler, LLB, Denise Dudzinski, PhD MTS, Ellen Fox, MD, Andrea Frolic, PhD, Jacqueline J. Glover, PhD, Kenneth Kipnis, PhD, Ann Marie Natali, MBA, William A. Nelson, PhD, Mary V. Rorty, PhD, Paul M. Schyve, MD, Joy D. Skeel, MDiv, and Anita J. Tarzian, PhD RN. *Core competencies for healthcare ethics consultation: the report of the American Society for Bioethics and Humanities.* 2nd Ed. (Glenview, IL: ASBH, American Society for Bioethics and Humanities, 2011), 2. Cf. Orlowski, James P., S. Hein, J. A. Christensen, R. Meinke, and Terry Sincich. "Why doctors use or do not use ethics consultation." *Journal of Medical Ethics* 32, no. 9 (2006): 499–503.

15. Arnold, Robert M., MD, Kenneth A. Berkowitz, MD, Nancy Neveloff Dubler, LLB, Denise Dudzinski, PhD MTS, Ellen Fox, MD, Andrea Frolic, PhD, Jacqueline J. Glover, PhD, Kenneth Kipnis, PhD, Ann Marie Natali, MBA, William A. Nelson, PhD, Mary V. Rorty, PhD, Paul M. Schyve, MD, Joy D. Skeel, MDiv, and Anita J. Tarzian, PhD RN. *Core competencies for healthcare ethics consultation: the report of the American Society for Bioethics and Humanities.* 2nd Ed. (Glenview, IL: ASBH, American Society for Bioethics and Humanities, 2011), 2.

16. Arnold, Robert M., MD, Kenneth A. Berkowitz, MD, Nancy Neveloff Dubler, LLB, Denise Dudzinski, PhD MTS, Ellen Fox, MD, Andrea Frolic, PhD, Jacqueline J. Glover, PhD, Kenneth Kipnis, PhD, Ann Marie Natali, MBA, William A. Nelson, PhD, Mary V. Rorty, PhD, Paul M. Schyve, MD, Joy D. Skeel, MDiv, and Anita J. Tarzian, PhD RN. *Core competencies for healthcare ethics consultation: the report of the American Society for Bioethics and Humanities.* 2nd Ed. (Glenview, IL: ASBH, American Society for Bioethics and Humanities, 2011), 2–3. cf; Oberle, Kathleen, and Dorothy Hughes, "Doctors' and nurses' perceptions of ethical problems in end-of-life decisions." *Journal of Advanced Nursing* 33, no. 6 (2001): 707–715.

17. Arnold, Robert M., MD, Kenneth A. Berkowitz, MD, Nancy Neveloff Dubler, LLB, Denise Dudzinski, PhD MTS, Ellen Fox, MD, Andrea Frolic, PhD, Jacqueline J. Glover, PhD, Kenneth Kipnis, PhD, Ann Marie Natali, MBA, William A. Nelson, PhD, Mary V. Rorty, PhD, Paul M. Schyve, MD, Joy D. Skeel, MDiv, and Anita J. Tarzian, PhD RN. *Core competencies for healthcare ethics consultation: the report of the American Society for Bioethics and Humanities*. 2nd Ed. (Glenview, IL: ASBH, American Society for Bioethics and Humanities, 2011), 2–3.

18. Arnold, Robert M., MD, Kenneth A. Berkowitz, MD, Nancy Neveloff Dubler, LLB, Denise Dudzinski, PhD MTS, Ellen Fox, MD, Andrea Frolic, PhD, Jacqueline J. Glover, PhD, Kenneth Kipnis, PhD, Ann Marie Natali, MBA, William A. Nelson, PhD, Mary V. Rorty, PhD, Paul M. Schyve, MD, Joy D. Skeel, MDiv, and Anita J. Tarzian, PhD RN. *Core competencies for healthcare ethics consultation: the report of the American Society for Bioethics and Humanities*. 2nd Ed. (Glenview, IL: ASBH, American Society for Bioethics and Humanities, 2011), 2–3.

19. Arnold, Robert M., MD, Kenneth A. Berkowitz, MD, Nancy Neveloff Dubler, LLB, Denise Dudzinski, PhD MTS, Ellen Fox, MD, Andrea Frolic, PhD, Jacqueline J. Glover, PhD, Kenneth Kipnis, PhD, Ann Marie Natali, MBA, William A. Nelson, PhD, Mary V. Rorty, PhD, Paul M. Schyve, MD, Joy D. Skeel, MDiv, and Anita J. Tarzian, PhD RN. *Core competencies for healthcare ethics consultation: the report of the American Society for Bioethics and Humanities*. 2nd Ed. (Glenview, IL: ASBH, American Society for Bioethics and Humanities, 2011), 2–3; cf. Childress, James F., Ruth R. Faden, Ruth D. Gaare, Lawrence O. Gostin, Jeffrey Kahn, Richard J. Bonnie, Nancy E. Kass, Anna C. Mastroianni, Jonathan D. Moreno, and Phillip Nieburg. "Public health ethics: mapping the terrain." *The Journal of Law, Medicine & Ethics* 30, no. 2 (2002): 170–178.

20. Post, Linda Farber., Jeffrey Blustein, and Nancy N. Dubler. *Handbook for Health Care Ethics Committees*, (Baltimore: Johns Hopkins University Press, 2007), 140–144.

21. Lo, Bernard, *Resolving Ethical Dilemmas: A Guide for Clinicians*, 4th ed (Philadelphia: Wolters Kluwer Health/Lippincott Williams & Wilkins, 2009), 67–69.

22. Orr, R. D., and W. Shelton. "A process and format for clinical ethics consultation." Journal of Clinical Ethics 20 no. 1, (2009): 79–80.

23. Orr, R. D., and W. Shelton. "A process and format for clinical ethics consultation." Journal of Clinical Ethics 20 no. 1, (2009): 80–82.

24. Post, Linda Farber., Jeffrey Blustein, and Nancy N. Dubler. *Handbook for Health Care Ethics Committees*, (Baltimore: Johns Hopkins University Press, 2007), 51–66.

25. Post, Linda Farber., Jeffrey Blustein, and Nancy N. Dubler. *Handbook for Health Care Ethics Committees*, (Baltimore: Johns Hopkins University Press, 2007), 30–36.

26. Lo, Bernard, *Resolving Ethical Dilemmas: A Guide for Clinicians*, 4th ed (Philadelphia: Wolters Kluwer Health/Lippincott Williams & Wilkins, 2009), 75–106.

27. Post, Linda Farber., Jeffrey Blustein, and Nancy N. Dubler. *Handbook for Health Care Ethics Committees*, (Baltimore: Johns Hopkins University Press, 2007), 67–80.

28. Orr, R. D., and W. Shelton. "A process and format for clinical ethics consultation." Journal of Clinical Ethics 20 no. 1, (2009): 80–83.

29. Orr, R. D., and W. Shelton. "A process and format for clinical ethics consultation." Journal of Clinical Ethics 20 no. 1, (2009): 80–82; *See* Shah, Seema, Amy Whittle, Benjamin Wilfond, Gary Gensler, and David Wendler. "How do institutional review boards apply the federal risk and benefit standards for pediatric research?." *JAMA* 291, no. 4 (2004): 476–482.

30. Orr, R. D., and W. Shelton. "A process and format for clinical ethics consultation." Journal of Clinical Ethics 20 no. 1, (2009): 82–85.

31. Orr, R. D., and W. Shelton. "A process and format for clinical ethics consultation." Journal of Clinical Ethics 20 no. 1, (2009): 82–85.

32. Orr, R. D., and W. Shelton. "A process and format for clinical ethics consultation." Journal of Clinical Ethics 20 no. 1, (2009): 82–88.

33. Staton, Jana, Roger W. Shuy, and Ira Byock, *A Few Months to Live: Different Paths to Life's End*, (Washington, D.C.: Georgetown University Press, 2001), 111–120.

34. Staton, Jana, Roger W. Shuy, and Ira Byock, *A Few Months to Live: Different Paths to Life's End*, (Washington, D.C.: Georgetown University Press, 2001), 123–130.
35. Orr, R. D., and W. Shelton. "A process and format for clinical ethics consultation." Journal of Clinical Ethics 20 no. 1, (2009): 82.
36. Orr, R. D., and W. Shelton. "A process and format for clinical ethics consultation." Journal of Clinical Ethics 20 no. 1, (2009): 82.
37. Orr, R. D., and W. Shelton. "A process and format for clinical ethics consultation." Journal of Clinical Ethics 20 no. 1, (2009): 82.
38. Orr, R. D., and W. Shelton. "A process and format for clinical ethics consultation." Journal of Clinical Ethics 20 no. 1, (2009): 82.
39. Orr, R. D., and W. Shelton. "A process and format for clinical ethics consultation." Journal of Clinical Ethics 20 no. 1, (2009): 82.
40. Fiester, Autumn. "Viewpoint: Why the Clinical Ethics We Teach Fails Patients," *Academci Medicine* 82, no 7 (2007): 684–689.
41. Jonsen, Albert R., Mark Siegler, and William J. Winslade. *Clinical Ethics: A Practical Approach to Ethical Decisions in Clinical Medicine*. 8th ed. (New York: McGraw Hill, Medical Pub. Division, 2015), 20–43.
42. Jonsen, Albert R., Mark Siegler, and William J. Winslade. *Clinical Ethics: A Practical Approach to Ethical Decisions in Clinical Medicine*. 8th ed. (New York: McGraw Hill, Medical Pub. Division, 2015), 93–101; *See* Flynn, Terry N., "Valuing citizen and patient preferences in health: recent developments in three types of best–worst scaling." *Expert review of pharmacoeconomics & outcomes research* 10, no. 3 (2010): 259–267.
43. Jonsen, Albert R., Mark Siegler, and William J. Winslade. *Clinical Ethics: A Practical Approach to Ethical Decisions in Clinical Medicine*. 8th ed. (New York: McGraw Hill, Medical Pub. Division, 2015), 20–32.
44. Jonsen, Albert R., Mark Siegler, and William J. Winslade. *Clinical Ethics: A Practical Approach to Ethical Decisions in Clinical Medicine*. 8th ed. (New York: McGraw Hill, Medical Pub. Division, 2015), 93–96.
45. Jonsen, Albert R., Mark Siegler, and William J. Winslade. *Clinical Ethics: A Practical Approach to Ethical Decisions in Clinical Medicine*. 8th ed. (New York: McGraw Hill, Medical Pub. Division, 2015), 12–13.
46. Jonsen, Albert R., Mark Siegler, and William J. Winslade. *Clinical Ethics: A Practical Approach to Ethical Decisions in Clinical Medicine*. 8th ed. (New York: McGraw Hill, Medical Pub. Division, 2015), 12–14.
47. Prager, Kenneth, MD, comp. "When Physicians and Surrogates Disagree about Futility." *American Medical Association Journal of Ethics* 15, no. 12 (December 2013): 1023.
48. Prager, Kenneth, MD, comp. "When Physicians and Surrogates Disagree about Futility." *American Medical Association Journal of Ethics* 15, no. 12 (December 2013): 1023.
49. Jonsen, Albert R., Mark Siegler, and William J. Winslade. *Clinical Ethics: A Practical Approach to Ethical Decisions in Clinical Medicine*. 8th ed. (New York: McGraw Hill, Medical Pub. Division, 2015), 93–96.
50. Jonsen, Albert R., Mark Siegler, and William J. Winslade. *Clinical Ethics: A Practical Approach to Ethical Decisions in Clinical Medicine*. 8th ed. (New York: McGraw Hill, Medical Pub. Division, 2015), 20–26.
51. Jonsen, Albert R., Mark Siegler, and William J. Winslade. *Clinical Ethics: A Practical Approach to Ethical Decisions in Clinical Medicine*. 8th ed. (New York: McGraw Hill, Medical Pub. Division, 2015), 49–54.
52. Jonsen, Albert R., Mark Siegler, and William J. Winslade. *Clinical Ethics: A Practical Approach to Ethical Decisions in Clinical Medicine*. 8th ed. (New York: McGraw Hill, Medical Pub. Division, 2015), 55–60.
53. Jonsen, Albert R., Mark Siegler, and William J. Winslade. *Clinical Ethics: A Practical Approach to Ethical Decisions in Clinical Medicine*. 8th ed. (New York: McGraw Hill, Medical Pub. Division, 2015), 93–101.

54. Jonsen, Albert R., Mark Siegler, and William J. Winslade. *Clinical Ethics: A Practical Approach to Ethical Decisions in Clinical Medicine*. 8th ed. (New York: McGraw Hill, Medical Pub. Division, 2015), 93–96; *See* Shalowitz, David I., Elizabeth Garrett-Mayer, and David Wendler. "The accuracy of surrogate decision makers: a systematic review." *Archives of internal medicine* 166, no. 5 (2006): 493–497.
55. Ganzini, L., L. Volicer, W. Nelson, E. Fox, and A. Derse. "Ten Myths About Decision-Making Capacity." *Journal of the American Medical Directors Association* 5, no.4 (2004): 263–267.
56. Jonsen, Albert R., Mark Siegler, and William J. Winslade. *Clinical Ethics: A Practical Approach to Ethical Decisions in Clinical Medicine*. 8th ed. (New York: McGraw Hill, Medical Pub. Division, 2015), 93–96.
57. Prager, Kenneth, MD, comp. "When Physicians and Surrogates Disagree about Futility." *American Medical Association Journal of Ethics* 15, no. 12 (December 2013): 1023.
58. Prager, Kenneth, MD, comp. "When Physicians and Surrogates Disagree about Futility." *American Medical Association Journal of Ethics* 15, no. 12 (December 2013): 1023.
59. Rubin, Susan B., *When Doctors Say No: The Battleground of Medical Futility*, (Bloomington, IN: Indiana University Press, 1998), 30–42.
60. onsen, Albert R., Mark Siegler, and William J. Winslade. *Clinical Ethics: A Practical Approach to Ethical Decisions in Clinical Medicine*. 8th ed. (New York: McGraw Hill, Medical Pub. Division, 2015), 112–148.
61. Jonsen, Albert R., Mark Siegler, and William J. Winslade. *Clinical Ethics: A Practical Approach to Ethical Decisions in Clinical Medicine*. 8th ed. (New York: McGraw Hill, Medical Pub. Division, 2015), 111.
62. Jonsen, Albert R., Mark Siegler, and William J. Winslade. *Clinical Ethics: A Practical Approach to Ethical Decisions in Clinical Medicine*. 8th ed. (New York: McGraw Hill, Medical Pub. Division, 2015), 69–84.
63. Staton, Jana, Roger W. Shuy, and Ira Byock, *A Few Months to Live: Different Paths to Life's End*, (Washington, D.C.: Georgetown University Press, 2001), 23–35.
64. Jonsen, Albert R., Mark Siegler, and William J. Winslade. *Clinical Ethics: A Practical Approach to Ethical Decisions in Clinical Medicine*. 8th ed. (New York: McGraw Hill, Medical Pub. Division, 2015), 165–175.
65. Jonsen, Albert R., Mark Siegler, and William J. Winslade. *ClinicalEthics: A Practical Approach to Ethical Decisions in Clinical Medicine*. 8th ed. (New York: McGraw Hill, Medical Pub. Division, 2015), 85–100.
66. Appelbaum, Paul S., Charles W. Lidz, and Alan Meisel, Informed Consent: Legal Theory and Clinical Practice, 2nd ed. (New York: Oxford University Press, 2001). 1.
67. Appelbaum, Paul S., Charles W. Lidz, and Alan Meisel, Informed Consent: Legal Theory and Clinical Practice, 2nd ed. (New York: OxfordUniversity Press, 2001). 11.
68. Jonsen, Albert R., Mark Siegler, and William J. Winslade. *Clinical Ethics: A Practical Approach to Ethical Decisions in Clinical Medicine*. 8th ed. (New York: McGraw Hill, Medical Pub. Division, 2015), 54–55.
69. ubin, Susan B., *When Doctors Say No: The Battleground of Medical Futility*, (Bloomington, IN: Indiana University Press, 1998), 35–42.
70. Jonsen, Albert R., Mark Siegler, and William J. Winslade. *Clinical Ethics: A Practical Approach to Ethical Decisions in Clinical Medicine*. 8th ed. (New York: McGraw Hill, Medical Pub. Division, 2015), 55–56; *See* Dowdy, Melvin D., Charles Robertson, and John A. Bander, "A study of proactive ethics consultation for critically and terminally ill patients with extended lengths of stay," *Critical care medicine* 26, no. 2 (1998): 252–259.
71. Jonsen, Albert R., Mark Siegler, and William J. Winslade. *Clinical Ethics: A Practical Approach to Ethical Decisions in Clinical Medicine*. 8th ed. (New York: McGraw Hill, Medical Pub. Division, 2015), 49–54.
72. Jonsen, Albert R., Mark Siegler, and William J. Winslade. *Clinical Ethics: A Practical Approach to Ethical Decisions in Clinical Medicine*. 8th ed. (New York: McGraw Hill, Medical Pub. Division, 2015), 55–57.

73. Post, Linda Farber., Jeffrey Blustein, and Nancy N. Dubler. *Handbook for Health Care Ethics Committees*, (Baltimore: Johns Hopkins University Press, 2007), 51–55.

74. Post, Linda Farber., Jeffrey Blustein, and Nancy N. Dubler. *Handbook for Health Care Ethics Committees*, (Baltimore: Johns Hopkins University Press, 2007), 177–180.

75. Jonsen, Albert R., Mark Siegler, and William J. Winslade. *Clinical Ethics: A Practical Approach to Ethical Decisions in Clinical Medicine*. 8th ed. (New York: McGraw Hill, Medical Pub. Division, 2015), 57.

76. Post, Linda Farber., Jeffrey Blustein, and Nancy N. Dubler. *Handbook for Health Care Ethics Committees*, (Baltimore: Johns Hopkins University Press, 2007), 96–98.

77. Jonsen, Albert R., Mark Siegler, and William J. Winslade. *Clinical Ethics: A Practical Approach to Ethical Decisions in Clinical Medicine*. 8th ed. (New York: McGraw Hill, Medical Pub. Division, 2015), 55.

78. Staton, Jana, Roger W. Shuy, and Ira Byock, *A Few Months to Live: Different Paths to Life's End*, (Washington, D.C.: Georgetown University Press, 2001), 100–125.

79. Jonsen, Albert R., Mark Siegler, and William J. Winslade. *Clinical Ethics: A Practical Approach to Ethical Decisions in Clinical Medicine*. 8th ed. (New York: McGraw Hill, Medical Pub. Division, 2015), 57; cf. Spatz, Erica S., Harlan M. Krumholz, and Benjamin W. Moulton. "The new era of informed consent: getting to a reasonable-patient standard through shared decision-making." *JAMA* 315, no. 19 (2016): 2063–2064.

80. Post, Linda Farber., Jeffrey Blustein, and Nancy N. Dubler. *Handbook for Health Care Ethics Committees*, (Baltimore: Johns Hopkins University Press, 2007), 19–23; cf. Wadlington, Walter. "Minors and health care: The age of consent." *Osgoode Hall LJ* 11 (1973): 115.

81. Jonsen, Albert R., Mark Siegler, and William J. Winslade. *Clinical Ethics: A Practical Approach to Ethical Decisions in Clinical Medicine*. 8th ed. (New York: McGraw Hill, Medical Pub. Division, 2015), 57.

82. Jonsen, Albert R., Mark Siegler, and William J. Winslade. *Clinical Ethics: A Practical Approach to Ethical Decisions in Clinical Medicine*. 8th ed. (New York: McGraw Hill, Medical Pub. Division, 2015), 57.

83. Staton, Jana, Roger W. Shuy, and Ira Byock, *A Few Months to Live: Different Paths to Life's End*, (Washington, D.C.: Georgetown University Press, 2001), 57–70.

84. Fletcher, Joseph F., *Situation Ethics: The New Morality*, (Philadelphia: Westminster Press, 1966), 17–25.

85. Oliver, Paul. "Conclusion: The Role of The Researcher." In *The Student's Guide to Research Ethics*. 2nd ed. Maidenhead, (Berkshire, England: McGraw-Hill/Open University Press, 2010), 171. *See* Tucker, Philip. *Good faith: in search of a unifying principle for the doctor-patient relationship*, (IA: Iowa State University Press, 1998), 13–40.

86. Oliver, Paul. "Conclusion: The Role of The Researcher." In *The Student's Guide to Research Ethics*. 2nd ed. Maidenhead, (Berkshire, England: McGraw-Hill/Open University Press, 2010), 171.

87. Oliver, Paul. "Conclusion: The Role of The Researcher." In *The Student's Guide to Research Ethics*. 2nd ed. Maidenhead, (Berkshire, England: McGraw-Hill/Open University Press, 2010), 171.

88. Fletcher, Joseph F., *Moral Responsibility: Situation Ethics at Work*, (Philadelphia: Westminster Press, 1967), 58–60.

89. Oliver, Paul. "Conclusion: The Role of The Researcher." In *The Student's Guide to Research Ethics*. 2nd ed. Maidenhead, (Berkshire, England: McGraw-Hill/Open University Press, 2010), 172.

90. Oliver, Paul. "Conclusion: The Role of The Researcher." In *The Student's Guide to Research Ethics*. 2nd ed. Maidenhead, (Berkshire, England: McGraw-Hill/Open University Press, 2010), 172.

91. Oliver, Paul. "Conclusion: The Role of The Researcher." In *The Student's Guide to Research Ethics*. 2nd ed. Maidenhead, (Berkshire, England: McGraw-Hill/Open University Press, 2010), 172.

92. Fletcher, Joseph F., *Moral Responsibility: Situation Ethics at Work*, (Philadelphia: Westminster Press, 1967), 29–40.
93. Fletcher, Joseph F., *Situation Ethics: The New Morality*, (Philadelphia: Westminster Press, 1966), 69–75.
94. Fletcher, Joseph F., *Moral Responsibility: Situation Ethics at Work*, (Philadelphia: Westminster Press, 1967), 60–65.
95. Fletcher, Joseph F., *Situation Ethics: The New Morality*, (Philadelphia: Westminster Press, 1966), 120–134.
96. Staton, Jana, Roger W. Shuy, and Ira Byock, *A Few Months to Live:Different Paths to Life's End*, (Washington, D.C.: Georgetown University Press, 2001), 57–89.
97. Fletcher, Joseph F., *Situation Ethics: The New Morality*, (Philadelphia: Westminster Press, 1966), 160–169.
98. Fox, Ellen, Kenneth A. Berkowitz, Barbara L. Chanko, and Tia Powell, "Ethics Consultation: Responding to Ethics Questions in Health Care." In *National Center for Ethics in Health Care*, (Washington: Veterans Health Administration, 2006), vi–vii.
99. Fox, Ellen, Kenneth A. Berkowitz, Barbara L. Chanko, and Tia Powell, "Ethics Consultation: Responding to Ethics Questions in Health Care." In *National Center for Ethics in Health Care*, (Washington: Veterans Health Administration, 2006), 27–31.
100. Fox, Ellen, Kenneth A. Berkowitz, Barbara L. Chanko, and Tia Powell, "Ethics Consultation: Responding to Ethics Questions in Health Care." In *National Center for Ethics in Health Care*, (Washington: Veterans Health Administration, 2006), 27–28.
101. Rubin, Susan B., *When Doctors Say No: The Battleground of Medical Futility*, (Bloomington, IN: Indiana University Press, 1998), 88–114. *See* Fried, Terri R., Elizabeth H. Bradley, Virginia R. Towle, and Heather Allore. "Understanding the treatment preferences of seriously ill patients," *New England Journal of Medicine* 346, no. 14 (2002): 1061–1066.
102. Fox, Ellen, Kenneth A. Berkowitz, Barbara L. Chanko, and Tia Powell, "Ethics Consultation: Responding to Ethics Questions in Health Care." In *National Center for Ethics in Health Care*, (Washington: Veterans Health Administration, 2006), 19–25.
103. Fox, Ellen, Kenneth A. Berkowitz, Barbara L. Chanko, and Tia Powell, "EthicsConsultation: Responding to Ethics Questions in Health Care." In *National Center for Ethics in Health Care*, (Washington: Veterans Health Administration, 2006), 32–44.
104. Fox, Ellen, Kenneth A. Berkowitz, Barbara L. Chanko, and Tia Powell, "EthicsConsultation: Responding to Ethics Questions in Health Care." In *National Center for Ethics in Health Care*, (Washington: Veterans Health Administration, 2006), 28–35.
105. Fox, Ellen, Kenneth A. Berkowitz, Barbara L. Chanko, and Tia Powell, "EthicsConsultation: Responding to Ethics Questions in Health Care." In *National Center for Ethics in Health Care*, (Washington: Veterans Health Administration, 2006), 20–27.
106. Fox, Ellen, Kenneth A. Berkowitz, Barbara L. Chanko, and Tia Powell, "Ethics Consultation: Responding to Ethics Questions in Health Care." In *National Center for Ethics in Health Care*, (Washington: Veterans Health Administration, 2006), 27–31.
107. Fox, Ellen, Kenneth A. Berkowitz, Barbara L. Chanko, and Tia Powell, "EthicsConsultation: Responding to Ethics Questions in Health Care." In *National Center for Ethics in Health Care*, (Washington: Veterans Health Administration, 2006), 44–46.
108. Fox, Ellen, Kenneth A. Berkowitz, Barbara L. Chanko, and Tia Powell, "Ethics Consultation: Responding to Ethics Questions in Health Care." In *National Center for Ethics in Health Care*, (Washington: Veterans Health Administration, 2006), 43–46. *See* Javier, Noel SC, and Marcos L. Montagnini, "Rehabilitation of the hospice and palliative care patient," *Journal of Palliative Medicine* 14, no. 5 (2011): 638–648.
109. Fox, Ellen, Kenneth A. Berkowitz, Barbara L. Chanko, and Tia Powell, "Ethics Consultation: Responding to Ethics Questions in Health Care." In *National Center for Ethics in Health Care*, (Washington: Veterans Health Administration, 2006), 47–49.
110. Antommaria Armond H., Berger Jeffrey, Berlinger Nancy, Carrese Joseph, Derse Art, Fiester Autumn, Fox Ellen, Gallagher Colleen M., Gallagher John, Goodman-Crews Paula,

Koogler Tracy, Latham Steve, Mitchell Christine, Mokwunye Nneka, Moskop John, Perlman Robert, Parsi Kayhan, Rosell Terry, Solomon Millie, Smith Martin, Spike Jeffery, Tarzian Anita, and Wocial Lucia, "CECA Report to the Board of Directors American Society for Bioethics and Humanities Certification, Accreditation, and Credentialing (C/A/C) of Clinical Ethics Consultants," (ASBH, 2010), 1–2 accessed at: http://asbh.org/uploads/about/CECA_Report_2010.pdf.

111. Antommaria Armond H., Berger Jeffrey, Berlinger Nancy, Carrese Joseph, Derse Art, Fiester Autumn, Fox Ellen, Gallagher Colleen M., Gallagher John, Goodman-Crews Paula, Koogler Tracy, Latham Steve, Mitchell Christine, Mokwunye Nneka, Moskop John, Perlman Robert, Parsi Kayhan, Rosell Terry, Solomon Millie, Smith Martin, Spike Jeffery, Tarzian Anita, and Wocial Lucia, "CECA Report to the Board of Directors American Society for Bioethics and Humanities Certification, Accreditation, and Credentialing (C/A/C) of Clinical Ethics Consultants," (ASBH, 2010), 2. accessed at: http://asbh.org/uploads/about/CECA_Report_2010.pdf

112. Antommaria Armond H., Berger Jeffrey, Berlinger Nancy, Carrese Joseph, Derse Art, Fiester Autumn, Fox Ellen, Gallagher Colleen M., Gallagher John, Goodman-Crews Paula, Koogler Tracy, Latham Steve, Mitchell Christine, Mokwunye Nneka, Moskop John, Perlman Robert, Parsi Kayhan, Rosell Terry, Solomon Millie, Smith Martin, Spike Jeffery, Tarzian Anita, and Wocial Lucia, "CECA Report to the Board of Directors American Society for Bioethics and Humanities Certification, Accreditation, and Credentialing (C/A/C) of Clinical Ethics Consultants," (ASBH, 2010), 2. accessed at: http://asbh.org/uploads/about/CECA_Report_2010.pdf

113. Antommaria Armond H., Berger Jeffrey, Berlinger Nancy, Carrese Joseph, Derse Art, Fiester Autumn, Fox Ellen, Gallagher Colleen M., Gallagher John, Goodman-Crews Paula, Koogler Tracy, Latham Steve, Mitchell Christine, Mokwunye Nneka, Moskop John, Perlman Robert, Parsi Kayhan, Rosell Terry, Solomon Millie, Smith Martin, Spike Jeffery, Tarzian Anita, and Wocial Lucia, "CECA Report to the Board of Directors American Society for Bioethics and Humanities Certification, Accreditation, and Credentialing (C/A/C) of Clinical Ethics Consultants," (ASBH, 2010), 2–3. accessed at: http://asbh.org/uploads/about/CECA_Report_2010.pdf

114. Antommaria Armond H., Berger Jeffrey, Berlinger Nancy, Carrese Joseph, Derse Art, Fiester Autumn, Fox Ellen, Gallagher Colleen M., Gallagher John, Goodman-Crews Paula, Koogler Tracy, Latham Steve, Mitchell Christine, Mokwunye Nneka, Moskop John, Perlman Robert, Parsi Kayhan, Rosell Terry, Solomon Millie, Smith Martin, Spike Jeffery, Tarzian Anita, and Wocial Lucia, "CECA Report to the Board of Directors American Society for Bioethics and Humanities Certification, Accreditation, and Credentialing (C/A/C) of Clinical Ethics Consultants," (ASBH, 2010), 3. accessed at: http://asbh.org/uploads/about/CECA_Report_2010.pdf

115. Antommaria Armond H., Berger Jeffrey, Berlinger Nancy, Carrese Joseph, Derse Art, Fiester Autumn, Fox Ellen, Gallagher Colleen M., Gallagher John, Goodman-Crews Paula, Koogler Tracy, Latham Steve, Mitchell Christine, Mokwunye Nneka, Moskop John, Perlman Robert, Parsi Kayhan, Rosell Terry, Solomon Millie, Smith Martin, Spike Jeffery, Tarzian Anita, and Wocial Lucia, "CECA Report to the Board of Directors American Society for Bioethics and Humanities Certification, Accreditation, and Credentialing (C/A/C) of Clinical Ethics Consultants," (ASBH, 2010), 2–4. accessed at: http://asbh.org/uploads/about/CECA_Report_2010.pdf

116. Antommaria Armond H., Berger Jeffrey, Berlinger Nancy, Carrese Joseph, Derse Art, Fiester Autumn, Fox Ellen, Gallagher Colleen M., Gallagher John, Goodman-Crews Paula, Koogler Tracy, Latham Steve, Mitchell Christine, Mokwunye Nneka, Moskop John, Perlman Robert, Parsi Kayhan, Rosell Terry, Solomon Millie, Smith Martin, Spike Jeffery, Tarzian Anita, and Wocial Lucia, "CECA Report to the Board of Directors American Society for Bioethics and Humanities Certification, Accreditation, and Credentialing (C/A/C) of Clinical

Ethics Consultants," (ASBH, 2010), 3–5. accessed at: http://asbh.org/uploads/about/CECA_Report_2010.pdf

117. Antommaria Armond H., Berger Jeffrey, Berlinger Nancy, Carrese Joseph, Derse Art, Fiester Autumn, Fox Ellen, Gallagher Colleen M., Gallagher John, Goodman-Crews Paula, Koogler Tracy, Latham Steve, Mitchell Christine, Mokwunye Nneka, Moskop John, Perlman Robert, Parsi Kayhan, Rosell Terry, Solomon Millie, Smith Martin, Spike Jeffery, Tarzian Anita, and Wocial Lucia, "CECA Report to the Board of Directors American Society for Bioethics and Humanities Certification, Accreditation, and Credentialing (C/A/C) of Clinical Ethics Consultants," (ASBH, 2010), 4. accessed at: http://asbh.org/uploads/about/CECA_Report_2010.pdf

118. Antommaria Armond H., Berger Jeffrey, Berlinger Nancy, Carrese Joseph, Derse Art, Fiester Autumn, Fox Ellen, Gallagher Colleen M., Gallagher John, Goodman-Crews Paula, Koogler Tracy, Latham Steve, Mitchell Christine, Mokwunye Nneka, Moskop John, Perlman Robert, Parsi Kayhan, Rosell Terry, Solomon Millie, Smith Martin, Spike Jeffery, Tarzian Anita, and Wocial Lucia, "CECA Report to the Board of Directors American Society for Bioethics and Humanities Certification, Accreditation, and Credentialing (C/A/C) of Clinical Ethics Consultants," (ASBH, 2010), 4. accessed at: http://asbh.org/uploads/about/CECA_Report_2010.pdf

119. Antommaria Armond H., Berger Jeffrey, Berlinger Nancy, Carrese Joseph, Derse Art, Fiester Autumn, Fox Ellen, Gallagher Colleen M., Gallagher John, Goodman-Crews Paula, Koogler Tracy, Latham Steve, Mitchell Christine, Mokwunye Nneka, Moskop John, Perlman Robert, Parsi Kayhan, Rosell Terry, Solomon Millie, Smith Martin, Spike Jeffery, Tarzian Anita, and Wocial Lucia, "CECA Report to the Board of Directors American Society for Bioethics and Humanities Certification, Accreditation, and Credentialing (C/A/C) of Clinical Ethics Consultants," (ASBH, 2010), 4. accessed at: http://asbh.org/uploads/about/CECA_Report_2010.pdf

120. Antommaria Armond H., Berger Jeffrey, Berlinger Nancy, Carrese Joseph, Derse Art, Fiester Autumn, Fox Ellen, Gallagher Colleen M., Gallagher John, Goodman-Crews Paula, Koogler Tracy, Latham Steve, Mitchell Christine, Mokwunye Nneka, Moskop John, Perlman Robert, Parsi Kayhan, Rosell Terry, Solomon Millie, Smith Martin, Spike Jeffery, Tarzian Anita, and Wocial Lucia, "CECA Report to the Board of Directors American Society for Bioethics and Humanities Certification, Accreditation, and Credentialing (C/A/C) of Clinical Ethics Consultants," (ASBH, 2010), 4–5. accessed at: http://asbh.org/uploads/about/CECA_Report_2010.pdf

121. Antommaria Armond H., Berger Jeffrey, Berlinger Nancy, Carrese Joseph, Derse Art, Fiester Autumn, Fox Ellen, Gallagher Colleen M., Gallagher John, Goodman-Crews Paula, Koogler Tracy, Latham Steve, Mitchell Christine, Mokwunye Nneka, Moskop John, Perlman Robert, Parsi Kayhan, Rosell Terry, Solomon Millie, Smith Martin, Spike Jeffery, Tarzian Anita, and Wocial Lucia, "CECA Report to the Board of Directors American Society for Bioethics and Humanities Certification, Accreditation, and Credentialing (C/A/C) of Clinical Ethics Consultants," (ASBH, 2010), 5. accessed at: http://asbh.org/uploads/about/CECA_Report_2010.pdf

122. Antommaria Armond H., Berger Jeffrey, Berlinger Nancy, Carrese Joseph, Derse Art, Fiester Autumn, Fox Ellen, Gallagher Colleen M., Gallagher John, Goodman-Crews Paula, Koogler Tracy, Latham Steve, Mitchell Christine, Mokwunye Nneka, Moskop John, Perlman Robert, Parsi Kayhan, Rosell Terry, Solomon Millie, Smith Martin, Spike Jeffery, Tarzian Anita, and Wocial Lucia, "CECA Report to the Board of Directors American Society for Bioethics and Humanities Certification, Accreditation, and Credentialing (C/A/C) of Clinical Ethics Consultants," (ASBH, 2010), 5. accessed at: http://asbh.org/uploads/about/CECA_Report_2010.pdf

123. Antommaria Armond H., Berger Jeffrey, Berlinger Nancy, Carrese Joseph, Derse Art, Fiester Autumn, Fox Ellen, Gallagher Colleen M., Gallagher John, Goodman-Crews Paula, Koogler Tracy, Latham Steve, Mitchell Christine, Mokwunye Nneka, Moskop John, Perlman

Robert, Parsi Kayhan, Rosell Terry, Solomon Millie, Smith Martin, Spike Jeffery, Tarzian Anita, and Wocial Lucia, "CECA Report to the Board of Directors American Society for Bioethics and Humanities Certification, Accreditation, and Credentialing (C/A/C) of Clinical Ethics Consultants," (ASBH, 2010), 5–6. accessed at: http://asbh.org/uploads/about/CECA_Report_2010.pdf

124. Antommaria Armond H., Berger Jeffrey, Berlinger Nancy, Carrese Joseph, Derse Art, Fiester Autumn, Fox Ellen, Gallagher Colleen M., Gallagher John, Goodman-Crews Paula, Koogler Tracy, Latham Steve, Mitchell Christine, Mokwunye Nneka, Moskop John, Perlman Robert, Parsi Kayhan, Rosell Terry, Solomon Millie, Smith Martin, Spike Jeffery, Tarzian Anita, and Wocial Lucia, "CECA Report to the Board of Directors American Society for Bioethics and Humanities Certification, Accreditation, and Credentialing (C/A/C) of Clinical Ethics Consultants," (ASBH, 2010), 5. accessed at: http://asbh.org/uploads/about/CECA_Report_2010.pdf

125. Antommaria Armond H., Berger Jeffrey, Berlinger Nancy, Carrese Joseph, Derse Art, Fiester Autumn, Fox Ellen, Gallagher Colleen M., Gallagher John, Goodman-Crews Paula, Koogler Tracy, Latham Steve, Mitchell Christine, Mokwunye Nneka, Moskop John, Perlman Robert, Parsi Kayhan, Rosell Terry, Solomon Millie, Smith Martin, Spike Jeffery, Tarzian Anita, and Wocial Lucia, "CECA Report to the Board of Directors American Society for Bioethics and Humanities Certification, Accreditation, and Credentialing (C/A/C) of Clinical Ethics Consultants," (ASBH, 2010), 5. accessed at: http://asbh.org/uploads/about/CECA_Report_2010.pdf

126. Antommaria Armond H., Berger Jeffrey, Berlinger Nancy, Carrese Joseph, Derse Art, Fiester Autumn, Fox Ellen, Gallagher Colleen M., Gallagher John, Goodman-Crews Paula, Koogler Tracy, Latham Steve, Mitchell Christine, Mokwunye Nneka, Moskop John, Perlman Robert, Parsi Kayhan, Rosell Terry, Solomon Millie, Smith Martin, Spike Jeffery, Tarzian Anita, and Wocial Lucia, "CECA Report to the Board of Directors American Society for Bioethics and Humanities Certification, Accreditation, and Credentialing (C/A/C) of Clinical Ethics Consultants," (ASBH, 2010), 5. accessed at: http://asbh.org/uploads/about/CECA_Report_2010.pdf

127. Antommaria Armond H., Berger Jeffrey, Berlinger Nancy, Carrese Joseph, Derse Art, Fiester Autumn, Fox Ellen, Gallagher Colleen M., Gallagher John, Goodman-Crews Paula, Koogler Tracy, Latham Steve, Mitchell Christine, Mokwunye Nneka, Moskop John, Perlman Robert, Parsi Kayhan, Rosell Terry, Solomon Millie, Smith Martin, Spike Jeffery, Tarzian Anita, and Wocial Lucia, "CECA Report to the Board of Directors American Society for Bioethics and Humanities Certification, Accreditation, and Credentialing (C/A/C) of Clinical Ethics Consultants," (ASBH, 2010), 5. accessed at: http://asbh.org/uploads/about/CECA_Report_2010.pdf

128. Antommaria Armond H., Berger Jeffrey, Berlinger Nancy, Carrese Joseph, Derse Art, Fiester Autumn, Fox Ellen, Gallagher Colleen M., Gallagher John, Goodman-Crews Paula, Koogler Tracy, Latham Steve, Mitchell Christine, Mokwunye Nneka, Moskop John, Perlman Robert, Parsi Kayhan, Rosell Terry, Solomon Millie, Smith Martin, Spike Jeffery, Tarzian Anita, and Wocial Lucia, "CECA Report to the Board of Directors American Society for Bioethics and Humanities Certification, Accreditation, and Credentialing (C/A/C) of Clinical Ethics Consultants," (ASBH, 2010), 5. accessed at: http://asbh.org/uploads/about/CECA_Report_2010.pdf

129. Antommaria Armond H., Berger Jeffrey, Berlinger Nancy, Carrese Joseph, Derse Art, Fiester Autumn, Fox Ellen, Gallagher Colleen M., Gallagher John, Goodman-Crews Paula, Koogler Tracy, Latham Steve, Mitchell Christine, Mokwunye Nneka, Moskop John, Perlman Robert, Parsi Kayhan, Rosell Terry, Solomon Millie, Smith Martin, Spike Jeffery, Tarzian Anita, and Wocial Lucia, "CECA Report to the Board of Directors American Society for Bioethics and Humanities Certification, Accreditation, and Credentialing (C/A/C) of Clinical Ethics Consultants," (ASBH, 2010), 5–6. accessed at: http://asbh.org/uploads/about/CECA_Report_2010.pdf

130. Anderson-Shaw, Lisa, DrPH MA MSN APN-C, Hannah Lipman, MD MS, Sally Bean, JD MA, D. Malcolm Shaner, MD, Courtenay R. Bruce, JD MA, Stuart Sprague, PhD, Stuart G. Finder, PhD, Lucia D. Wocial, PhD RN, and Aviva Katz, MD MA FCAS FAAP, eds. *Improving Competencies in Clinical Ethics Consultation: An Education Guide,* 2nd ed. (Chicago, Illinois: American Society for Bioethics and Humanities, 2015), 2–4.
131. Antommaria Armond H., Berger Jeffrey, Berlinger Nancy, Carrese Joseph, Derse Art, Fiester Autumn, Fox Ellen, Gallagher Colleen M., Gallagher John, Goodman-Crews Paula, Koogler Tracy, Latham Steve, Mitchell Christine, Mokwunye Nneka, Moskop John, Perlman Robert, Parsi Kayhan, Rosell Terry, Solomon Millie, Smith Martin, Spike Jeffery, Tarzian Anita, and Wocial Lucia, "CECA Report to the Board of Directors American Society for Bioethics and Humanities Certification, Accreditation, and Credentialing (C/A/C) of Clinical Ethics Consultants," (ASBH, 2010), 6. accessed at: http://asbh.org/uploads/about/CECA_Report_2010.pdf
132. Antommaria Armond H., Berger Jeffrey, Berlinger Nancy, Carrese Joseph, Derse Art, Fiester Autumn, Fox Ellen, Gallagher Colleen M., Gallagher John, Goodman-Crews Paula, Koogler Tracy, Latham Steve, Mitchell Christine, Mokwunye Nneka, Moskop John, Perlman Robert, Parsi Kayhan, Rosell Terry, Solomon Millie, Smith Martin, Spike Jeffery, Tarzian Anita, and Wocial Lucia, "CECA Report to the Board of Directors American Society for Bioethics and Humanities Certification, Accreditation, and Credentialing (C/A/C) of Clinical Ethics Consultants," (ASBH, 2010), 6. accessed at: http://asbh.org/uploads/about/CECA_Report_2010.pdf
133. Antommaria Armond H., Berger Jeffrey, Berlinger Nancy, Carrese Joseph, Derse Art, Fiester Autumn, Fox Ellen, Gallagher Colleen M., Gallagher John, Goodman-Crews Paula, Koogler Tracy, Latham Steve, Mitchell Christine, Mokwunye Nneka, Moskop John, Perlman Robert, Parsi Kayhan, Rosell Terry, Solomon Millie, Smith Martin, Spike Jeffery, Tarzian Anita, and Wocial Lucia, "CECA Report to the Board of Directors American Society for Bioethics and Humanities Certification, Accreditation, and Credentialing (C/A/C) of Clinical Ethics Consultants," (ASBH, 2010), 6. accessed at: http://asbh.org/uploads/about/CECA_Report_2010.pdf
134. Antommaria Armond H., Berger Jeffrey, Berlinger Nancy, Carrese Joseph, Derse Art, Fiester Autumn, Fox Ellen, Gallagher Colleen M., Gallagher John, Goodman-Crews Paula, Koogler Tracy, Latham Steve, Mitchell Christine, Mokwunye Nneka, Moskop John, Perlman Robert, Parsi Kayhan, Rosell Terry, Solomon Millie, Smith Martin, Spike Jeffery, Tarzian Anita, and Wocial Lucia, "CECA Report to the Board of Directors American Society for Bioethics and Humanities Certification, Accreditation, and Credentialing (C/A/C) of Clinical Ethics Consultants," (ASBH, 2010), 8. accessed at: http://asbh.org/uploads/about/CECA_Report_2010.pdf
135. Antommaria Armond H., Berger Jeffrey, Berlinger Nancy, Carrese Joseph, Derse Art, Fiester Autumn, Fox Ellen, Gallagher Colleen M., Gallagher John, Goodman-Crews Paula, Koogler Tracy, Latham Steve, Mitchell Christine, Mokwunye Nneka, Moskop John, Perlman Robert, Parsi Kayhan, Rosell Terry, Solomon Millie, Smith Martin, Spike Jeffery, Tarzian Anita, and Wocial Lucia, "CECA Report to the Board of Directors American Society for Bioethics and Humanities Certification, Accreditation, and Credentialing (C/A/C) of Clinical Ethics Consultants," (ASBH, 2010), 8–9. accessed at: http://asbh.org/uploads/about/CECA_Report_2010.pdf
136. Antommaria Armond H., Berger Jeffrey, Berlinger Nancy, Carrese Joseph, Derse Art, Fiester Autumn, Fox Ellen, Gallagher Colleen M., Gallagher John, Goodman-Crews Paula, Koogler Tracy, Latham Steve, Mitchell Christine, Mokwunye Nneka, Moskop John, Perlman Robert, Parsi Kayhan, Rosell Terry, Solomon Millie, Smith Martin, Spike Jeffery, Tarzian Anita, and Wocial Lucia, "CECA Report to the Board of Directors American Society for Bioethics and Humanities Certification, Accreditation, and Credentialing (C/A/C) of Clinical Ethics Consultants," (ASBH, 2010), 8–9. accessed at: http://asbh.org/uploads/about/CECA_Report_2010.pdf

137. Antommaria Armond H., Berger Jeffrey, Berlinger Nancy, Carrese Joseph, Derse Art, Fiester Autumn, Fox Ellen, Gallagher Colleen M., Gallagher John, Goodman-Crews Paula, Koogler Tracy, Latham Steve, Mitchell Christine, Mokwunye Nneka, Moskop John, Perlman Robert, Parsi Kayhan, Rosell Terry, Solomon Millie, Smith Martin, Spike Jeffery, Tarzian Anita, and Wocial Lucia, "CECA Report to the Board of Directors American Society for Bioethics and Humanities Certification, Accreditation, and Credentialing (C/A/C) of Clinical Ethics Consultants," (ASBH, 2010), 9. accessed at: http://asbh.org/uploads/about/CECA_Report_2010.pdf

138. Antommaria Armond H., Berger Jeffrey, Berlinger Nancy, Carrese Joseph, Derse Art, Fiester Autumn, Fox Ellen, Gallagher Colleen M., Gallagher John, Goodman-Crews Paula, Koogler Tracy, Latham Steve, Mitchell Christine, Mokwunye Nneka, Moskop John, Perlman Robert, Parsi Kayhan, Rosell Terry, Solomon Millie, Smith Martin, Spike Jeffery, Tarzian Anita, and Wocial Lucia, "CECA Report to the Board of Directors American Society for Bioethics and Humanities Certification, Accreditation, and Credentialing (C/A/C) of Clinical Ethics Consultants," (ASBH, 2010), 8–9.

139. Antommaria Armond H., Berger Jeffrey, Berlinger Nancy, Carrese Joseph, Derse Art, Fiester Autumn, Fox Ellen, Gallagher Colleen M., Gallagher John, Goodman-Crews Paula, Koogler Tracy, Latham Steve, Mitchell Christine, Mokwunye Nneka, Moskop John, Perlman Robert, Parsi Kayhan, Rosell Terry, Solomon Millie, Smith Martin, Spike Jeffery, Tarzian Anita, and Wocial Lucia, "CECA Report to the Board of Directors American Society for Bioethics and Humanities Certification, Accreditation, and Credentialing (C/A/C) of Clinical Ethics Consultants," (ASBH, 2010), 8. accessed at: http://asbh.org/uploads/about/CECA_Report_2010.pdf

140. Antommaria Armond H., Berger Jeffrey, Berlinger Nancy, Carrese Joseph, Derse Art, Fiester Autumn, Fox Ellen, Gallagher Colleen M., Gallagher John, Goodman-Crews Paula, Koogler Tracy, Latham Steve, Mitchell Christine, Mokwunye Nneka, Moskop John, Perlman Robert, Parsi Kayhan, Rosell Terry, Solomon Millie, Smith Martin, Spike Jeffery, Tarzian Anita, and Wocial Lucia, "CECA Report to the Board of Directors American Society for Bioethics and Humanities Certification, Accreditation, and Credentialing (C/A/C) of Clinical Ethics Consultants," (ASBH, 2010), 9. accessed at: http://asbh.org/uploads/about/CECA_Report_2010.pdf

141. Antommaria Armond H., Berger Jeffrey, Berlinger Nancy, Carrese Joseph, Derse Art, Fiester Autumn, Fox Ellen, Gallagher Colleen M., Gallagher John, Goodman-Crews Paula, Koogler Tracy, Latham Steve, Mitchell Christine, Mokwunye Nneka, Moskop John, Perlman Robert, Parsi Kayhan, Rosell Terry, Solomon Millie, Smith Martin, Spike Jeffery, Tarzian Anita, and Wocial Lucia, "CECA Report to the Board of Directors American Society for Bioethics and Humanities Certification, Accreditation, and Credentialing (C/A/C) of Clinical Ethics Consultants," (ASBH, 2010), 9. accessed at: http://asbh.org/uploads/about/CECA_Report_2010.pdf

142. Antommaria Armond H., Berger Jeffrey, Berlinger Nancy, Carrese Joseph, Derse Art, Fiester Autumn, Fox Ellen, Gallagher Colleen M., Gallagher John, Goodman-Crews Paula, Koogler Tracy, Latham Steve, Mitchell Christine, Mokwunye Nneka, Moskop John, Perlman Robert, Parsi Kayhan, Rosell Terry, Solomon Millie, Smith Martin, Spike Jeffery, Tarzian Anita, and Wocial Lucia, "CECA Report to the Board of Directors American Society for Bioethics and Humanities Certification, Accreditation, and Credentialing (C/A/C) of Clinical Ethics Consultants," (ASBH, 2010), 9. accessed at: http://asbh.org/uploads/about/CECA_Report_2010.pdf

143. Antommaria Armond H., Berger Jeffrey, Berlinger Nancy, Carrese Joseph, Derse Art, Fiester Autumn, Fox Ellen, Gallagher Colleen M., Gallagher John, Goodman-Crews Paula, Koogler Tracy, Latham Steve, Mitchell Christine, Mokwunye Nneka, Moskop John, Perlman Robert, Parsi Kayhan, Rosell Terry, Solomon Millie, Smith Martin, Spike Jeffery, Tarzian Anita, and Wocial Lucia, "CECA Report to the Board of Directors American Society for Bioethics and Humanities Certification, Accreditation, and Credentialing (C/A/C) of Clinical

Ethics Consultants," (ASBH, 2010), 9. accessed at: http://asbh.org/uploads/about/CECA_Report_2010.pdf

144. Antommaria Armond H., Berger Jeffrey, Berlinger Nancy, Carrese Joseph, Derse Art, Fiester Autumn, Fox Ellen, Gallagher Colleen M., Gallagher John, Goodman-Crews Paula, Koogler Tracy, Latham Steve, Mitchell Christine, Mokwunye Nneka, Moskop John, Perlman Robert, Parsi Kayhan, Rosell Terry, Solomon Millie, Smith Martin, Spike Jeffery, Tarzian Anita, and Wocial Lucia, "CECA Report to the Board of Directors American Society for Bioethics and Humanities Certification, Accreditation, and Credentialing (C/A/C) of Clinical Ethics Consultants," (ASBH, 2010), 9. accessed at: http://asbh.org/uploads/about/CECA_Report_2010.pdf

145. Larson, Eric B. "Clinical Empathy as Emotional Labor in the Patient-Physician Relationship." *JAMA* 293, no. 9, (2005): 40–45.

146. Antommaria Armond H., Berger Jeffrey, Berlinger Nancy, Carrese Joseph, Derse Art, Fiester Autumn, Fox Ellen, Gallagher Colleen M., Gallagher John, Goodman-Crews Paula, Koogler Tracy, Latham Steve, Mitchell Christine, Mokwunye Nneka, Moskop John, Perlman Robert, Parsi Kayhan, Rosell Terry, Solomon Millie, Smith Martin, Spike Jeffery, Tarzian Anita, and Wocial Lucia, "CECA Report to the Board of Directors American Society for Bioethics and Humanities Certification, Accreditation, and Credentialing (C/A/C) of Clinical Ethics Consultants," (ASBH, 2010), 10. accessed at: http://asbh.org/uploads/about/CECA_Report_2010.pdf

147. Antommaria Armond H., Berger Jeffrey, Berlinger Nancy, Carrese Joseph, Derse Art, Fiester Autumn, Fox Ellen, Gallagher Colleen M., Gallagher John, Goodman-Crews Paula, Koogler Tracy, Latham Steve, Mitchell Christine, Mokwunye Nneka, Moskop John, Perlman Robert, Parsi Kayhan, Rosell Terry, Solomon Millie, Smith Martin, Spike Jeffery, Tarzian Anita, and Wocial Lucia, "CECA Report to the Board of Directors American Society for Bioethics and Humanities Certification, Accreditation, and Credentialing (C/A/C) of Clinical Ethics Consultants," (ASBH, 2010), 10. accessed at: http://asbh.org/uploads/about/CECA_Report_2010.pdf

148. Antommaria Armond H., Berger Jeffrey, Berlinger Nancy, Carrese Joseph, Derse Art, Fiester Autumn, Fox Ellen, Gallagher Colleen M., Gallagher John, Goodman-Crews Paula, Koogler Tracy, Latham Steve, Mitchell Christine, Mokwunyc Nneka, Moskop John, Perlman Robert, Parsi Kayhan, Rosell Terry, Solomon Millie, Smith Martin, Spike Jeffery, Tarzian Anita, and Wocial Lucia, "CECA Report to the Board of Directors American Society for Bioethics and Humanities Certification, Accreditation, and Credentialing (C/A/C) of Clinical Ethics Consultants," (ASBH, 2010), 10. accessed at: http://asbh.org/uploads/about/CECA_Report_2010.pdf

149. Antommaria Armond H., Berger Jeffrey, Berlinger Nancy, Carrese Joseph, Derse Art, Fiester Autumn, Fox Ellen, Gallagher Colleen M., Gallagher John, Goodman-Crews Paula, Koogler Tracy, Latham Steve, Mitchell Christine, Mokwunye Nneka, Moskop John, Perlman Robert, Parsi Kayhan, Rosell Terry, Solomon Millie, Smith Martin, Spike Jeffery, Tarzian Anita, and Wocial Lucia, "CECA Report to the Board of Directors American Society for Bioethics and Humanities Certification, Accreditation, and Credentialing (C/A/C) of Clinical Ethics Consultants," (ASBH, 2010), 10. accessed at: http://asbh.org/uploads/about/CECA_Report_2010.pdf

150. Antommaria Armond H., Berger Jeffrey, Berlinger Nancy, Carrese Joseph, Derse Art, Fiester Autumn, Fox Ellen, Gallagher Colleen M., Gallagher John, Goodman-Crews Paula, Koogler Tracy, Latham Steve, Mitchell Christine, Mokwunye Nneka, Moskop John, Perlman Robert, Parsi Kayhan, Rosell Terry, Solomon Millie, Smith Martin, Spike Jeffery, Tarzian Anita, and Wocial Lucia, "CECA Report to the Board of Directors American Society for Bioethics and Humanities Certification, Accreditation, and Credentialing (C/A/C) of Clinical Ethics Consultants," (ASBH, 2010), 10–11. accessed at: http://asbh.org/uploads/about/CECA_Report_2010.pdf

151. Antommaria Armond H., Berger Jeffrey, Berlinger Nancy, Carrese Joseph, Derse Art, Fiester Autumn, Fox Ellen, Gallagher Colleen M., Gallagher John, Goodman-Crews Paula, Koogler Tracy, Latham Steve, Mitchell Christine, Mokwunye Nneka, Moskop John, Perlman Robert, Parsi Kayhan, Rosell Terry, Solomon Millie, Smith Martin, Spike Jeffery, Tarzian Anita, and Wocial Lucia, "CECA Report to the Board of Directors American Society for Bioethics and Humanities Certification, Accreditation, and Credentialing (C/A/C) of Clinical Ethics Consultants," (ASBH, 2010), 10–11. accessed at: http://asbh.org/uploads/about/CECA_Report_2010.pdf

152. Antommaria Armond H., Berger Jeffrey, Berlinger Nancy, Carrese Joseph, Derse Art, Fiester Autumn, Fox Ellen, Gallagher Colleen M., Gallagher John, Goodman-Crews Paula, Koogler Tracy, Latham Steve, Mitchell Christine, Mokwunye Nneka, Moskop John, Perlman Robert, Parsi Kayhan, Rosell Terry, Solomon Millie, Smith Martin, Spike Jeffery, Tarzian Anita, and Wocial Lucia, "CECA Report to the Board of Directors American Society for Bioethics and Humanities Certification, Accreditation, and Credentialing (C/A/C) of Clinical Ethics Consultants," (ASBH, 2010), 11. accessed at: http://asbh.org/uploads/about/CECA_Report_2010.pdf

153. Antommaria Armond H., Berger Jeffrey, Berlinger Nancy, Carrese Joseph, Derse Art, Fiester Autumn, Fox Ellen, Gallagher Colleen M., Gallagher John, Goodman-Crews Paula, Koogler Tracy, Latham Steve, Mitchell Christine, Mokwunye Nneka, Moskop John, Perlman Robert, Parsi Kayhan, Rosell Terry, Solomon Millie, Smith Martin, Spike Jeffery, Tarzian Anita, and Wocial Lucia, "CECA Report to the Board of Directors American Society for Bioethics and Humanities Certification, Accreditation, and Credentialing (C/A/C) of Clinical Ethics Consultants," (ASBH, 2010), 11. accessed at: http://asbh.org/uploads/about/CECA_Report_2010.pdf

154. Antommaria Armond H., Berger Jeffrey, Berlinger Nancy, Carrese Joseph, Derse Art, Fiester Autumn, Fox Ellen, Gallagher Colleen M., Gallagher John, Goodman-Crews Paula, Koogler Tracy, Latham Steve, Mitchell Christine, Mokwunye Nneka, Moskop John, Perlman Robert, Parsi Kayhan, Rosell Terry, Solomon Millie, Smith Martin, Spike Jeffery, Tarzian Anita, and Wocial Lucia, "CECA Report to the Board of Directors American Society for Bioethics and Humanities Certification, Accreditation, and Credentialing (C/A/C) of Clinical Ethics Consultants," (ASBH, 2010), 11. accessed at: http://asbh.org/uploads/about/CECA_Report_2010.pdf

155. Antommaria Armond H., Berger Jeffrey, Berlinger Nancy, Carrese Joseph, Derse Art, Fiester Autumn, Fox Ellen, Gallagher Colleen M., Gallagher John, Goodman-Crews Paula, Koogler Tracy, Latham Steve, Mitchell Christine, Mokwunye Nneka, Moskop John, Perlman Robert, Parsi Kayhan, Rosell Terry, Solomon Millie, Smith Martin, Spike Jeffery, Tarzian Anita, and Wocial Lucia, "CECA Report to the Board of Directors American Society for Bioethics and Humanities Certification, Accreditation, and Credentialing (C/A/C) of Clinical Ethics Consultants," (ASBH, 2010), 11. accessed at: http://asbh.org/uploads/about/CECA_Report_2010.pdf

156. Prichard, Harold Arthur. Moral Obligation and Duty and Interest: Essays and Lectures. London: Oxford U.P, 1968.

157. Antommaria Armond H., Berger Jeffrey, Berlinger Nancy, Carrese Joseph, Derse Art, Fiester Autumn, Fox Ellen, Gallagher Colleen M., Gallagher John, Goodman-Crews Paula, Koogler Tracy, Latham Steve, Mitchell Christine, Mokwunye Nneka, Moskop John, Perlman Robert, Parsi Kayhan, Rosell Terry, Solomon Millie, Smith Martin, Spike Jeffery, Tarzian Anita, and Wocial Lucia, "CECA Report to the Board of Directors American Society for Bioethics and Humanities Certification, Accreditation, and Credentialing (C/A/C) of Clinical Ethics Consultants," (ASBH, 2010), 12. accessed at: http://asbh.org/uploads/about/CECA_Report_2010.pdf

158. Antommaria Armond H., Berger Jeffrey, Berlinger Nancy, Carrese Joseph, Derse Art, Fiester Autumn, Fox Ellen, Gallagher Colleen M., Gallagher John, Goodman-Crews Paula, Koogler Tracy, Latham Steve, Mitchell Christine, Mokwunye Nneka, Moskop John, Perlman Robert,

Parsi Kayhan, Rosell Terry, Solomon Millie, Smith Martin, Spike Jeffery, Tarzian Anita, and Wocial Lucia, "CECA Report to the Board of Directors American Society for Bioethics and Humanities Certification, Accreditation, and Credentialing (C/A/C) of Clinical Ethics Consultants," (ASBH, 2010), 12--13. accessed at: http://asbh.org/uploads/about/CECA_Report_2010.pdf

159. Antommaria Armond H., Berger Jeffrey, Berlinger Nancy, Carrese Joseph, Derse Art, Fiester Autumn, Fox Ellen, Gallagher Colleen M., Gallagher John, Goodman-Crews Paula, Koogler Tracy, Latham Steve, Mitchell Christine, Mokwunye Nneka, Moskop John, Perlman Robert, Parsi Kayhan, Rosell Terry, Solomon Millie, Smith Martin, Spike Jeffery, Tarzian Anita, and Wocial Lucia, "CECA Report to the Board of Directors American Society for Bioethics and Humanities Certification, Accreditation, and Credentialing (C/A/C) of Clinical Ethics Consultants," (ASBH, 2010), 12–14. accessed at: http://asbh.org/uploads/about/CECA_Report_2010.pdf

160. Fox, Ellen, Kenneth A. Berkowitz, Barbara L. Chanko, and Tia Powell, "Ethics Consultation: Responding to Ethics Questions in Health Care." In *National Center for Ethics in Health Care*, (Washington: Veterans Health Administration, 2006), 47–49. accessed at: http://asbh.org/uploads/about/CECA_Report_2010.pdf

# Chapter 3
# Moral Reasoning, Ethics Facilitation and Virtue

Exploring a basis for analytic moral reasoning becomes tenable when a philosophical foundation for moral judgments and ethical facilitation evolves into a competency-based curriculum. Analytic moral reasoning in health care ethics consultation is highly beneficial because it expects and encourages ethicists to engage with dilemmas critically [1]. Alternatively, consultants tend to observe the current skills of health care ethics consultations dogmatically without analytic moral reasoning [2]. Dogmatic interpretation of rules creates rigidity and does not allow for adaptation according to the variance of ethical situations [3]. A standard set of rules and evaluative methods for clinical ethics consultation, although practical, lacks the plasticity needed to adapt to ethical issues in health care. Reimagining standard clinical ethics consultation approaches as a philosophical purist into praxis encourages critical thinking and bolsters current consultation methods [4].

Contemporary ethics largely bisects into deontological and utilitarian approaches. These options are popular among normative theoristsbecause they focus on a *good* end. Typically, the end is a moral agent or group of moral agents. Situational variances in medicine provoke pause amongst professionals when a course of action is unclear or disputed. Good moral philosophy underlies the spectrum of medical discrepancies but applying moral philosophy under acceptable circumstances is the ultimate challenge in training clinical ethicists [5].

Rather than the clinical situation, moral emphasis on the agentpromotes examining an individual's traits rather than the moral obligations dictated by normative ethics. However, virtue ethics serves as a combatant viewpoint that counters the rigidity of deontology and utilitarian pursuits. Typically, virtue ethics is an approach that diverges with deontology'sseminal thinker, Immanuel Kant. Contemporary moral theorists like Alasdair MacIntyre and Philippa Foot have chastised Kant as a thinker who rigidly structures morality into a process of rule-following and obedience [6]. In many respects, Kant contributes to analytic philosophy's rejection of virtue. These criticisms poke at the shortcomings Kant's moral theory displays when juxtaposed with traditional Greek virtue ethics insofar as Kant's deontological

J. T. Bertino, *Clinical Ethics for Consultation Practice*, https://doi.org/10.1007/978-3-030-90182-0_3

approach heavily bases itself upon an ethic of dogmatic rules rather than an ethic of virtue [7].The criticisms of deontological thinking should not discourage ethicists from understanding the value of Kant's moral theory and its interrelatedness with virtue ethics [8].

Virtue ethics attempts to identify the agent as a morally "good" person and subsequently determine what the good agent would do or would not do. This understanding of virtue ethics prioritizes ontology over praxis. In terms of Kantian virtue ethics, acts that entail ends in themselves rather than a means to an end are ethically formidable due to the moral motivation of the agent. When the agent's motivations direct toward a good that directs or intends a good action, the agent acts virtuously [9]. The Kantian moral agent is the quintessence of a good health care ethics consultant. If a consultant approaches an ethical situation with the same good that will encompass a morally fortuitous agent, the consultant acts consistently and respects the rules and protocols of the hospital, state, and any other governing body [10].

Rule governing principles like the four pillars of bioethics are primarily associated with Kant's theory of moral behavior. The maxims associated with Kant's deontological structure underlie a moral agent's intentions. These maxims pertain to honorable intentions that apply to the long-term goals of patients or individual decision-makers. Since long-term goals become tenable through various means, the maxims relevant to health care ethics consultants relate to ends in themselves [11]. Kant explains this argument in his *Metaphysics of Morals* by explaining that all acts have ended and that ends are objects of free choice. Following this argument, the ends toward which an agent directs their actions are self-governing.

Consequently, the maxims an agent constructs are also self-governed insofar as these maxims direct toward a moral end. However, two moral ends that take precedence in Kant's moral theory are the agent's duty to strive toward perfection and promote happiness in others. These two ends are ultimate responsibilities in terms of achieving virtue [12].Kant's moral theory emphasizes virtue at its core through obligation and duty. Nonetheless, the two ends which moral agents ought to dutifully pursue must receive further analysis. Health care ethics consultants must critically analyze situations, facilitate discussion, and possess skills for identifying and evaluating moral judgments analytically.

Among the skills ethics consultants must acquire, analytic moral reasoning is a category of imaginative skills that do not typically receive attention. The ASBH has given little attention to this area precisely because of its seemingly inapplicable nature. The practical skills that ethicists learn and exercise typically involve proper bedside manner, informal mediation, and rules that dictate permissibility. However, by examining a philosophical basis for virtue, ethicists may become moral agents in their practice and subsequently engage in rational moral decision-making when exercising practical consultation skills. A moral agent accomplishes the specific ends that the agent must pursue by exercising virtues that inevitably engage in ends themselves. In other words, the virtues are ends in themselves rather than means to various ends. For the sake of this analysis, the virtues of wisdom, justice, compassion, and humility serve as expressive skills that promote a consultant's analytic moral reasoning abilities.

It is fitting, beginning with the virtue of wisdom due to the relevance this virtue possesses in honing one's ability to analytically reason. Since ethics consultation is an interpretive endeavor, necessary experience and knowledge areprominent components. Wisdom allows health care ethicists to ascertain details of a moral situation. Ranging from medical indications to patient preferences, the virtuous ethicist cannot be satisfied with minimum descriptions of a clinical situation [13]. Instead, the ethicists must evaluate the observable facts and comprehend a detailed account of the events and interactions. The level of detail, although initially robust, subsequently reduces to ascertainable pieces of information that emphasize depth over breadth [14]. Still, a significant gap in a clinical ethicist's knowledge is the medical indications and the clinical circumstances that dictate ethical uncertainties.

Commonly referenced in health care ethics, justice stands among four focal virtues associated with principlism. Understanding justice in moral virtue directs a consultant's end toward bettering themselves, their organization, and their patients. Understanding the virtue of justice allows ethicists to maneuver between health care ethics structures that may or may not yield benefits for patients [15]. Justice promotes analytical reasoning by encouraging ethicists to use a sensible approach to threatening patients and their self-flourishing. Among these various issues are gender discrimination, resource allocation, and hospital hierarchies. Rather than just dissecting a situation's intricacies into visual elements of right and wrong, emphasizing justice as a virtue in health care ethics encourages analytic reasoning for consultants and aids facilitation [16].

While justice embodies the epitome of Kantian ends by progressing moral agents toward the betterment of self and others, courage serves as a virtue that combats some of the most severe issues in bioethics. Courage possesses tremendous influence in developing analytical reasoning skills because of its ability to promote a moral agent's beliefs [17]. Since the benefit of others and self-improvement are the two leading Kantian ends toward which ethics must direct itself, courage aids analytical reasoning by challenging moral agents to adhere to their beliefs and what is ethically appropriate in the context of the consultation.

The final two relevant virtues for developing analytic moral reasoning skills in health care ethics consultation are compassion and humility. These virtues hold specific relevance toward the betterment of others through empathetic introspection [18]. Ethics consultants can easily slip into emotive notions when encountering a patient under physical and emotional duress. Although an ethicist's emotions ought to bolster the ethicist's abilities to reflect on human suffering, emotions should never cloud or disrupt the ethicist's focus and decision-making abilities [19]. Still, the inherent bias that accompanies the vocational nature of clinical ethics will forever pose a gap between ethics consultants and other stakeholders.

Consultation approaches often overlook humility in both moral virtue theory and consultation methods. Humility requires ethicists to approach each medical situation with a humble attitude and a reserved demeanor. The knowledge an ethicist possesses is often overwhelming for patients, families, and health care professionals. Due to the sensitivity of ethical situations, patients and involved stakeholders may become disillusioned by complex information or attitudes that exert arrogance.

Considering this, ethicists should acknowledge their abilities and bracket their expertise. Ethics consultants should access a clinical case's essential and relevant informationwhen encountering a situation that demands ethical expertise [20].

Wisdom, courage, justice, compassion, and humility are virtues that a health care ethics consultant must identify and practice if the consultant exercises analytic moral reasoning skills. Although various virtues may also apply to this analysis, they are primary bases for analytic reasoning because of their roots in Kantian virtue theory. By first understanding the primary ends towards which a consultant must gear his attention, all subsequent ways an ethicist conducts a consultation may direct the betterment of himself and the patient [21]. Although these virtues receive priority in this analysis, a key element in honing analytic reasoning skills interconnects with one's ability to identify virtue. Moreover, identifying virtue is a task that one cannot easily accomplish since a virtuous life is an activity that contains ends within its practice. The pursuit of virtue is never complete and requires perpetual refinement.

While Kant's conception of virtue outlines an epistemic demand, the ontological identification of virtue still requires consideration. Furthermore, identifying virtue in a health care ethics consultation allows for practical application in health care ethics consultations. Moral theorists often attribute virtue identification to Aristotelian virtue ethics. In doing so, Aristotle's ethics uncovers practical ways in which an ethicist may achieve analytic moral reasoning skills in his practice [22].

Virtue serves as ends in themselves that inherently directs one's functions toward the betterment of oneself and the people affected by one's actions as the agent. However, identifying how each virtue manifests becomes difficult without proper guidance. Identifying virtue must become the primary objective of a health care ethics consultant if the consultant aspires to perform his function adequately. Nonetheless, identifying each virtue in terms of analytic moral reasoning is extremely difficult in health care ethics due to the severity of the situations involved and the array of implicated stakeholders. By addressing an Aristotelian approach to identifying virtue and the subsequent actions that follow, specifically the development of analytic moral reasoning skills, the critical components of a clinical ethics consultant's curriculum will demonstrate necessary learning points and describe traits that a consultant must possess.

Aristotle notably establishes a mean or average, ethical approach for identifying virtue. One of his primary tasks in his *Nicomachean Ethics* is determining what behaviors yield happiness and how happiness may become an achievable goal. Aristotle claims that happiness requires performing a function well. For health care ethics consultants, performing the primary functions of facilitation and conflict management are paramount. These activities, though admittedly onerous, are tenable if consultants use analytic moral reasoning skills through the medium of virtue. Happiness for self and others occurs by living virtuously and performing one's function well. Performing one's function well entails directing one's moral ends toward self-improvement and the improvement of others [23].

Identifying virtue in a practical sense first requires a moral agent's willingness to become self-directing toward beneficial ends. Without the desire to improve and perform one's function well, a clinical ethics consultant cannot recognize how virtue manifests practically and effectively [24]. Virtues are complex and often misidentified due to desires, thoughts, images, emotions, and actions. Nevertheless, the appetitive desires associated with dissuading virtue diminish once the moral agent accepts his function as a facilitator and resource for patients, families, and health care professionals [25].Kant's ethical assessment posits that emotions are irrational impulses that guide a moral agent's reasoning abilities. However, Kant also notes that emotion takes shape through moral education. While moral education begins early, reasoning through complex moral situations in health care requires the perpetual drilling of concepts and skills rooted in virtue [26]. Since this analysis aims to demonstrate how clinical ethics consultants may develop heightened analytic moral reasoning skills, a method that aids ethicists in identifying their appetitive desires and emotions both promotes poise within the moral agent and serves as a means of identifying virtue in ethical practice [27].

Adopting an Aristotelian approach to identifying virtue catalyzes one's efforts to live virtuously. These processes are mutually inclusive in Aristotle's ethics. Aristotle attempts to uncover the intermediate nature of virtues (1105b25–6). This section of Aristotle's ethics, commonly referred to as the "Aristotelian Doctrine of the Mean," typically understands moral agency as a duty to identify and recognize the intermediary understanding of emotional regulation. Properly assessing and regulating emotions exposes a corresponding virtue as a median point [28]. For instance, courage is a virtue that lies between cowardice and rashness. In contrast, a deficiency in courage manifests in cowardice; an excess of courage results in impulsive and often unreasonable action (1106a26–b28) [29].

The Aristotelian doctrine of the mean possesses applicability to all virtues, yet identifying virtues still present issues. Identifying a mean for any situation does not necessarily yield virtue at the end of the proverbial rainbow. Rather than exclusively adopt the Kantian ends toward which moral agents should direct their actions, a reimagining of Aristotle's doctrine of the mean proves beneficial in health care situations. The varying emotions that occur while performing a health care ethics consultation may make or break a formidable ethical assessment. Although various practical skills aid the consultation process, these skills cannot become practical without first developing moral analytical reasoning skills [30]. Identifying and honing analytic moral reasoning skills is a tenable activity if the moral agent can first recognize his emotions regarding a situation and subsequently implement warranted emotions when necessary. Aristotle's ethics insists that moral agents attempt to use certain emotions at the correct times regarding situations, involved stakeholders, and motives. In this respect, the same appetitive emotions that dissuade analytical moral reasoning may become beneficial tools. For instance, appropriate anger is beneficial when fueling a moral agent's desire for courage, during reasonable fear and confidence aid temperance [31].

The approach to identifying virtue mentioned above in no way condones the exile of emotion, for emotion is a necessary component to conducting successful ethical assessments. By way of example, compassion must accompany emotion in some capacity because it is a virtue that allows one to empathize with another's suffering. If a moral agent engages in compassion, he must identify with the sufferer in some way. In many respects, compassion requires the moral agent to see himself as a vulnerable community member, i.e., the sufferer. Finally, compassion requires an appetitive desire to help. If the moral agent seeks to alleviate the suffering of an individual, they must inevitably feel for the weakness of the suffering individual [32]. With this assessment in mind, moral agents must emote in their endeavors to live virtuously. However, identifying and using proper emotions only becomes possible if a moral agent seeks to exercise his profession virtuously. It follows that analytic moral reasoning skills develop when a moral agent identifies the proper use of emotions in ethical situations to perform a function virtuously [33].

Although it may be clear that emotion is a necessary component for identifying compassion, emotion is still relevant across all virtues. Courage requires a proper allocation of warranted emotions due to its rootedness in fear. Without fear, courageous action cannot occur because the moral agent does not endure a difficult situation. In the Platonic dialogue the *Laches,* Socrates indicates that deliberation, calculation, and reason are possible through wisdom. To endure through a frightening situation is ultimately empty unless the fear itself is subordinate to wisdom [34]. Health care ethics consultants can learn a great deal from Plato's teachings. Fear accompanies difficult ethical situations in health care for all parties involved. Despite this, if the moral agent regulates emotions properly by performing a function well, the moral agent identifies virtue and exercises analytical moral reasoning skills [35].

Identifying virtue and honing subsequent moral reasoning skills allow health care ethics consultants to use these skills in their practice. Although the variance of ethical situations in health care makes consultation efforts difficult, regulating the fervent desires of patients, families, and even consultants is a difficult task. Well-developed virtue ethics occurs when negotiation occurs. Without proper negotiation between involved individuals, relationships will not flourish. Proper regulation of emotions may come differently to various consultants. By way of example, presenting the practical facts may provoke emotion for one ethicist but not for another. For the sake of this analysis, good analytic moral assessment comes from entering a situation with the understanding that proper emotive action must become regulated through an understanding of virtue. Virtue identification becomes especially relevant under Aristotelian virtue ethics [36].

Aristotle states that the moral agent must know that he must perform virtuous actions, decide upon these actions, and perform these actions from a firm and unchanging disposition (1105a30-1105b) [37]. These three points do not specify that emotion cannot accompany the moral agent at the decision-making outset. Instead, having an emotional background before assessing the situation, provided the moral agent performs proper emotional regulation, may provide dutiful notions of right and wrong [38]. This assessment also aligns with principlism by appealing to the goals of the four principles of bioethics. Nonmaleficence's principle is especially pertinent to this discussion as it accompanies a fervent desire to do intentional harm to patients. While the principle is clear, the emotion that accompanies the principle absconds from the agent'saccompanying principle applies to a situation.

Appealing to virtue is more effective than other consultation means because analytic moral guidance emerges from analytic moral reasoning [39].

A modern contribution by moral theologian Joseph Fletcher aids this analysis by providing an insightful method of using virtue ethics in contemporary ethical practice. Through examining Fletcher's method, the demand for analytic moral reasoning skills in health care ethics consultations becomes apparent. Furthermore, due to his theory's ability to adapt to current ethical infrastructures, Fletcher's pragmatic approach to virtue ethics justifies and augments the ASBH's practical skills listed above. Fletcher is known for developing situation ethics—an ethical approach that understands how different ethical situations may manifest [40]. Although the vast differences between human beings and their clinical situations bolster the beauty of autonomy, an ethical approach that can adapt to the differences of human beings is significantly under virtue. Situation ethics does not limit its practice to a singularity and thus does not apply standards and ethical norms to all situations [41].

In some respect, Fletcher's approach demands a great deal of heroism from clinical ethicists for various reasons. Fletcher's analysis insists that an ethicist must regulate his emotion using virtue identification, performing his duties virtuously, and subsequently engaging in analytic moral reasoning skills to maximize facilitation efforts. However, this task becomes exceptionally onerous in the face of terror, fear, or a determination for self-protection. The heroism of an ethicist derives from performing his duties in the face of fear with nothing but analytic moral reasoning at his side [42]. Analogous with the Aristotelian soldier, a clinical ethicist must regulate his emotions by implementing analytic moral reasoning. Rushing into a situation that presents objective danger or risk to a patient or involved stakeholder is rash, while decisional stagnancy provoked by fear indicates cowardice. Alternatively, if the ethicist uses proper emotive regulation by implementing moral reasoning skills into his decision, he makes a balanced recommendation that embodies patience and effective facilitation [43].

Traditional conceptions of virtue and moral theory possess qualities that can contribute to clinical ethics consultation throughout the philosophical corpus. Kant's deontological principles, Aristotle's doctrine of the mean, Fletcher's situation ethics, and the pragmatic approaches of Beauchamp and Childress' principlism may have a crucial hand in modifying ethics consultation competencies and developing bioethics a professional discipline [44]. Additionally, the ASBH's core skills align with the ethical goals of the moral theorists listed above due to the cooperative nature of professional ethics. However, while governing bodies that regulate the educational facets of ethics consultation and teaching have adopted various methods in illustrating relevant information for clinical ethics consultants, key factors regarding philosophical notions of character development, stewardship, integrity, and virtue have become diminished in the educational curriculum for bioethicists [45]. Understandably, professionals responsible for designing ethics competencies and curriculums may deem theseirrelevant or perhaps too broad for ethics education [46]. Aristotle himself asserts that no ethical theory can apply to a decisional procedure insofar as no set of ordinary rules can ever identify various emotions like anger (1125b26–7) [47].

## The ASBH and Core Competencies

In 1998, the ASBH revealed the first edition of the *Core Competencies for Health Care Ethics Consultation*. This report was the first of its kind in many respects [48]. At the time, the core competencies spawned from a project that required the efforts of various professionals and health care organizations. The conglomerate of twenty-one individuals possessed a like-minded approach to clinical ethics in that the task force focused on defining the nature and goals of clinical ethics consultation, identifying the types of knowledge and skills involved in consultation efforts, addressing organizational ethics issues, and discussing the importance of consultations under the guise of specific institutional policies [49]. Furthermore, the task force responsible for developing a set of core competencies for budding clinical ethicists further reflected its passion and dedication for clinical ethics by incorporating the expertise of a diverse group of individuals who contribute to the multi-disciplinary nature of bioethics. The Core Competencies intend to establish a methodology for assessing medical ethics issues. While the core focus of the text is to prepare individuals to conduct clinical ethics consultations, the project's unique character arises with an understanding that no formal project's unique characteris the ASBH's inception of the core competencies [50].

The demanding task that the ASBH's task force set out to complete culminated in a well-constructed and formatted piece of literature covering a tremendous amount of information with brevity and clarity. Naturally, developing technologies spark further moral questions in medicine. The nature and scope of consultation competencies must adapt and remain malleable throughout these developments [51]. Emerging technologies, cultural changes, and hospital protocols all affect the scope and design of a set of competencies that intend to benefit clinical ethics consultants. Due to the chronic shift in clinical focus, the ASBH released a second edition of the core competencies in 2011. This text included expansions and amendments of previous sections that seemingly improved the overall competencies and, in the eyes of the task force, better encompassed the necessary information required of clinical ethics consultants [52].

Although the second edition of the ASBH's core competencies amended its focus and scope due to healthcare's changing nature, the new ASBH task force removed a critical dimension of moral reasoning from the original edition. The second component of the current version of the ASBH's core competencies, "Core Competencies for Health Care Ethics Consultation," marks a significant replacement. Subsection 2.4, "Attributes, Attitudes, and Behaviors of Ethics Consultants," replaces the original title, "Character and Ethics Consultation." [53] To compensate for removing a character component to ethics consultation, the second edition of the ASBH's core competencies includes a brief presentation on "Moral Reasoning and Ethical Theory." This section appears under the "Core Knowledge for HCEC" section of the text and includes four ethical perspectives: Consequentialist/non-consequentialist approaches, theological/religious approaches, principlism, and related theories of justice [54].

While the changes between editions seem small, the detriments of these changes lie in the removal of competencies about the underlying elements of an excellent clinical ethics consultant [55]. Although presented as simply "character" in the first edition of the text, these elements adhere to the regulation of virtue in clinical ethics consultations [56]. The underlying shortcoming of the ASBH's amendments lies in a focus shift. This shift begins with an emphasis on philosophical concepts and migrates to a behavioral perspective. Despite using virtue and subsequent analytic moral reasoning skills unique to this analysis, the ASBH's removal of character and substitution with behavior/attitude components creates a fundamental gap in the curriculum that otherwise may have bolstered moral reasoning skills. While the first edition of the ASBH's curriculum possessed one half of a vital teaching component, the second edition removed this half and replaced it with another. In effect, a formidable ethics curriculum between both editions becomes possible if the analysis satisfies two conditions. First, ethicists must categorize both character and behavioral functions, or emotive functions. Second, clinical ethicists' character and behavioral functions are core skills, not knowledge points surrounding the discipline. In doing so, moral development becomes efficient and effective by introducing skills that regulate decision-making and facilitate character development during clinical recommendations [57]. The regulation of emotion inevitably enhances proper character development. As a result, clinical ethics consultants gain analytic moral reasoning skills to identify virtue, deficiencies, and excesses [58].

Character development is an essential aspect of clinical ethics consultation. While the practical skills and core knowledge outlined throughout the ASBH core competencies covers an extensive amount of information, the matter of character opens various pathways to virtuous behaviors and attitudes that permeate beyond the confines of clinical ethics consultation and saturate the health care organizations. The connection between character and organizational ethics could also explain the amendment of section three in the first edition of the ASBH's core competencies, "Organizational Ethics," along with its subsections: "Defining Organizational Ethics" and "Some Preliminary Recommendations." [59] It is no coincidence that the piece on character precedes the piece on organizational ethics. This additional amendment further illustrates the importance of character in health care ethics consultation and the areas of health care it affects [60].

Addressing the importance of character in the ASBH's core competencies demonstrates the value of virtueethics in shaping a moral atmosphere and aids in illustrating the nature of a competency [61]. The role of a clinical ethics consultant is an important one due to the impact a consultant's recommendations possess. The vastness of a clinical ethics consultant's expertise requires justification within its practice. The concept of competencies allows consultants to remain diligent in their work and aids consultants in thinking quickly and efficiently [62]. Clinical situations move quickly, and so too must ethicists. Just as is the case for professional care staff, an ethicist cannot pause and retreat to a library to assess an ethical theory or find aid in justifying a clinical decision. Instead, competencies must serve as training tools and educational points that leave a lasting impression on professional ethics consultants [63]. Nonetheless, insofar as character promotes moral virtue and reasoning, the character's role is a vital tool for ensuring a consultant's ability to exercise and implement competencies properly [64].

Expanding on the need for character development in clinical ethics consultation requires an analysis of the scope and function of skill competencies. In doing so, skill competencies present themselves as educational tools that demonstrate their applicability to various facets of a health care organization, including chaplaincy, social work, and lay-person occupations [65].

By elaborating the ASBH's understanding of core competency skills, the need and utility of character development for clinical ethics consultants become crucial for bioethicists. Additionally, the goal of uncovering and implementing analytic moral reasoning skills throughout a consultation system becomes a far more tenable feat if true character regulates and promotes balanced emotions [66]. Emotions are inseparable and fundamental aspects of decision-making in health care. Nevertheless, emotions may result in poor decision-making if the moral agent does not properly regulate his emotions during ethical deliberations [67]. Whilompetencies provide practical skills regarding the occupation of health care ethics consultation and how one conducts consultations, these skills risk becoming misunderstood, misused, and inappropriately interpreted if the competencies lack proper emotive guidance [68].

The examination of the nature and function of emotion in decision-making processes grants a greater understanding of competency development and curriculum formation. Furthermore, the demand and need for emotion regulation become vital within clinical ethics practice. While taming emotions is possible, the regulatory catalyst of virtue ethics serves as a teachable model for shaping character. The omission of character development in the ASBH's second edition of the core competencies undermines the nature of clinical ethics consultation and inhibits the proper emotive development of clinical decision-makers [69]. Nonetheless, both editions of the ASBH's core competencies present fundamental components that allude to the importance of virtue identification. Inevitably, virtue identification techniques shape emotion and subsequently foster analytic moral reasoning skills for ethics consultants. These skills improve the overall quality of consultations and pragmatically direct clinical bioethics toward a promising future in professional health care [70].

The ethically relevant information that clinical ethics consultants must demonstrate a great area across a health care institution's infrastructure and permeates medical situations that range from the social to the terminal. By establishing competencies in consultation, ethicists become reaffirmed in their expertise and responsibilities [71]. The rationale for competencies in health care ethics consultations remains unchanged. Under its goal of quality improvement, the ASBH task force intends to educate and guide consultants through the variety of ethical discrepancies that form in clinical and organizational situations [72]. While emotional regulation does not receive attention, competency skills receive tremendous priority due to their effectiveness [73].

The competencies intend to expose the strengths and weaknesses of consultants productively. Be that as it may, the ASBH task force on ethics competencies addresses and compares the methods in which competencies apply in different mediums: individual consultants, consultation teams, referral services, off-site services, and ethics committees [74]. While each consultation medium possesses advantages and disadvantages, the concept of character development through emotional regulation, virtue-identification techniques, and analytic moral reasoning skills remain pertinent and beneficial aspects to clinical ethics consultation [75].

The priority of the ASBH's core skills lies in the fact that without these skills, the following knowledge points outlined in the ASBH's consultation curriculum cannot apply to clinical situations. The skills divide into three sets: ethical assessment skills, process skills, and interpersonal skills. Beginning with ethical assessment skills, identifying the nature of a conflict or ethical discrepancy is the first step in determining the need for a consultation. While the consultant is not necessarily responsible for calling a consultation, he is responsible for gathering the relevant information needed to assess and recommend options for involved stakeholders. Some of the process skills involved with ethical assessments include access to medically relevant information, recognizing the social and interpersonal dimensions of involved stakeholders, and evaluating one's limitations and involvement with a specific case [76].

Process skills also contain various dimensions that risk unsuccessful responses to an ethics consultation request due to unregulated emotion. Process skills include an ethicist's ability to understand and relay the realistic expectations of a given clinical situation. These skills involve identifying which individuals need to become involved with a consultation and the kind of consultation medium that would most effectively address a situation, *i.e.,* committee, individual, external service, etc. [77] While process skills encompass necessary aspects of two clinical ethics consultations, these skills are also subject to emotive sway if not adequately regulated. Process skills involve communicative and collaborative efforts to work with other professionals, stakeholders, and patients effectively. Avoiding bias by regulating emotions well-ordered increases patient safety measures and aids in proper communication between involved departments. However, these skills also require character development through analytic moral reasoning, resulting in fatal errors and ethical shortcomings [78].

Finally, and perhaps most relevant in terms of character development, including interpersonal skills. Clinical ethics consultants must conduct social interactions with involved stakeholders for various reasons. Understanding a clinical situation requires a consultant to listen well and communicate their interest, respect, and support. Recognizing the relationship barriers between stakeholders aids facilitation efforts and adds an educational component to the consultation process [79]. Nevertheless, a lack of moral fortitude and character may seriously damage one's interpersonal skillset for various reasons. First, consultants who cannot relate to a clinical situation in a balanced and regulated manner risk engaging in an excess or deficiency of ethical practice compassionately. The resulting outcome can seriously compromise professional recommendations due to a lack of regulated emotion [80]. Second, without a formidable and developed character and analytic moral reasoning skills, consultants may be unable to educate individuals about their options and seriously compromise learning opportunities for professionals, patients, and other staff members. Third, a lack of analytic moral reasoning via poor character development results in skewed views and perceptions about a clinical situation. While a consultant must facilitate and resolve conflict, this shortcoming can result in the opposite, namely, provoke discrepancies between involved stakeholders [81].

The ethical assessment, process, and interpersonal skills outlined by the ASBH core competencies are essential skills that possess tremendous practicality and efficacy. However, without the proper regulation of a clinical ethicist's emotions, the facilitation of situations through a virtuous lens hinders conflict-resolution efforts and thus diminishes the use of analytic moral reasoning [82]. Alternatively, the development of analytic moral reasoning skills through moral discernment aids emotional regulation [83].

The purpose of the competency skills illustrated above captures the effectiveness of a well-formulated ethics curriculum. The skills and knowledge listed in both editions of the ASBH's core competencies are extensive and cover a tremendous amount of material that possesses real applicability in clinical ethics situations. While the knowledge points include various facets of health care ethics that a consultant should become familiar with, *i.e.,* patient rights, principle-based reasoning, end-of-life decision-making options, genetic testing, and counseling, etc., the skills portion of the competencies is a far more critical prerequisite set of information for clinical ethics consultants [84].

The three focal topics of this analysis include emotion, virtue, and analytic moral reasoning. This analysis intends to demonstrate analytic moral reasoning as a set of competency skills. However, emotion and virtue are necessary components that aid in developing a moral agent's ability to hone these skills [85]. While this analysis stresses the importance of identifying virtue in clinical practice, the practicality and function of this task may be brutal for individuals who lack a formal background in analytic moral theory. Despite this dilemma, bioethics is a multi-disciplinary field that welcomes varying expertise. By examining instances of virtue identification, analytic moral reasoning skills emerge and demonstrate their effectiveness in contemporary clinical ethics curriculums [86].

Honing analytic moral reasoning skills requires practice and engagement. While clinical ethicists possess various skills, their ability to reason to adopt both practical and efficient methods required in professional health care presents difficulties. For instance, the ASBH's core competencies are a well-received text because it possesses practical skills that teach clinicians practical methods in health care ethics consultation. Be that as it may, the practical skills outlined by the ASBH become far less effective if moral agents cannot deliberate effectively. Proper deliberation includes mutual reasoning abilities on both sides. Virtue prototypically manifests as an abstract concept that rarely receives praise for its propensity for practical application. By identifying virtue in practical instances, the analytical and the theoretical conjoin, inevitably shaping a teachable curriculum for ethicists.

Examining the philosophical foundations of informed consent in medical research by addressing the philosophical theories of deontological and utilitarian ethics illuminates alternative theories and exposes the problems that accompany them. Thus far, this assessmentattempts to promote a turn toward virtue ethics as a practical means of obtaining mutual understanding between involved stakeholders in health care. One topic in health care that encapsulates this dichotomy is research ethics. The discussion between researchers and research participants must receive proper attention due to the contentious and controversial history of medical research on human subjects.

## Ethics and Research

Due to its power, influence, and lasting impression on contemporary medical research with humans, this historical analysis of informed consent begins with the Nuremberg Code. Although vastly specialized toward the crimes committed during World War II, the ten-point statement of the Nuremberg Code dramatically contributes to a formalized understanding of informed consent in research. It delineates a foundational understanding of human rights when performing medical research [87]. This examination of the Nuremberg Code's historical influence will explain the foundational genesis of informed consent in medical research and allow for an examination of the problems that accompany a lack of informed consent and the philosophical justifications for informed consent.

The one-hundred thirty-nine-day Nuremberg trial exposed the war crimes against humanity committed in third Reich concentration camps experiments during World War II. The trial judges based those accused on foundational philosophical principles that demonstrated a massive violation of human rights, bioethical principles, and notions of humanity [88]. The Nuremberg Code demonstrated its uniqueness by emphasizing principles of natural law and human rights concerning medical experimentation with humans [89]. Furthermore, the Nuremberg Code's philosophical points attempted to articulate the importance and meaning behind informed consent. Informed consent matters are at the heart of the various crimes committed against human beings during the Nazi experiments. Thus, the trials attempted to formulate informed consent and implement informed consent in clinical practice and research [90].

Despite the comprehensiveness of the Nuremberg Code, it only addresses issues of informed consent that were especially relevant to the crimes committed during the Nazi experiments. The rules and regulations detailed in the code receive criticism as more applicable to barbaric individuals than civilized individuals [91]. However, contrary to these criticisms, the Nuremberg Code sparked a worldwide consensus regarding the voluntariness involved in obtaining consent, especially when experimentation with human subjects [92]. Furthermore, the Nuremberg Code's indication of the importance of obtaining free and voluntary informed consent reinforces philosophical ideals in place when performing any kind of intervention with human beings.

These philosophical principles are reinforced in the ten points the Nuremberg Code outlines. The first point stresses the overall importance of the document and the goal it attempts to accomplish by emphasizing the absolute necessity for the voluntary consent of human subjects [93]. The first point stipulates that informed consent requires uncoerced participation from subjects [94]. The first point also discusses the importance of providing research participants with ample information before participating in the study [95]. The following points detail the importance and philosophical implications of informed consent in greater detail by specifying nuances accompanying the Nazi experiments. These points include the importance of a study contributing to the good of society, protections for patients by first running trials on animals thus ensuring the experiment is safe, and the right of informed patients to discontinue their participation in a study if they find the study is detrimental to their health and safety [96].

Despite the importance and influence of the Nuremberg code, some consider the document incomplete concerning its guiding role in ethical research [97].For instance, the document does not consider pediatric research, vulnerable populations, or mentally impaired individuals [98]. Nevertheless, the document emphasizes informed consent and applies various philosophical principles that accompany informed consent, albeit esoterically. The two relevant philosophical principles this analysis addresses in terms of informed consent include deontology and utilitarianism. Nevertheless, to examine the philosophical theories implicated by informed consent, this analysis must analyze issues that informed consent addresses and the underlying justifications [99].

The Nuremberg trials and subsequent code exposed the horrific medical experiments during World War II and prompted biomedical ethics to emphasize informed consent in medical research [100]. Although informed consent is at the forefront of the Nuremberg trials' ethical discussion, the following code begs the question: What makes informed consent so vastly crucial in medical research with humans? The justification for informed consent lies in the principle of autonomy. Autonomy, or self-care, describes the inherent rights of human beings to freely make decisions that, in concurrence with their discretion, serve as a means for choosing justly because the agent freely makes the decision [101]. The Nuremberg Code exemplifies the need to uphold autonomy in its stipulations that state that another individual cannot have dominion and control another human being [102]. Autonomy involves promoting liberty and agency. Liberty entails independent choosing that is free of influence, while agency entails the capacity and capability of an individual to choose [103]. The mandatory respect for autonomy is rooted in the fact that rational human beings possess the capacity to choose and make decisions for themselves. Autonomy in informed consent justifies moral norms that, when followed, allow further investigation into standards of medical practice. The Nuremberg Code and the autonomous standards it attempts to uphold inherently include two integral aspects to moral-ethical practice, namely, nonmaleficence and beneficence.

Tom Beauchamp and James Childress focused their understanding of harm as the inhibition of one's flourishing [104]. Harm, additionally, is especially pertinent to inducing pain, death, disability, or suffering [105]. Nonmaleficence, as a normative obligation in research ethics, specifies five rules that aid in ethical decision-making. These rules include: (1) do not kill; (2) do not cause pain or suffering; (3) do not incapacitate; (4) do not cause offense; and (5) do not deprive others of the goods of life [106]. These rules emphasize the safety, dignity, respect, and interest of a human being. Since autonomy is the inherent self-care and reasoning that human beings possess, there are inherent moral obligations researchers must abide by when conducting human research [107].

The principles of nonmaleficence and beneficence serve as philosophical foundations of ethical practice. William Frankena addresses these points by combining nonmaleficence and beneficence into a single set of theories. Frankena presents four obligations: (1) One ought not to inflict evil or harm; (2) one ought to prevent evil or harm; (3) one ought to remove evil or harm; and (4) one ought to do or promote well [108]. The first obligation is one of nonmaleficence, while the remaining three refer to obligations of beneficence [109].Frankena's combination of obligations presents another example of normative ethical principles and rules established to justify research with human beings.

The obligations presented by Frankena emphasize the importance of limiting or eliminating possible harm that may befall research participants [110]. Be that as it may, Frankena's combined theories result in a set of norms that may categorize under the umbrella of deontological ethics. Although effective in some capacity, the deontological aspects of fixed principles can become too stringent and provoke rigidity when developing a moral theory that remains effective across various cases. This analysis examines two distinct philosophical positions that apply to contemporary medical research. These positions include deontology and utilitarianism.

The development and history of informed consent aid this analysis by introducing the importance of acquiring consent in medical research with humans and the philosophical import and foundational basis autonomy serve when performing medical research on human beings. The lack of autonomous choice that implicitly results in not acquiring informed consent presents various philosophical issues. Nevertheless, since the Nuremberg Code's inception, various philosophical theories have been implemented and used to justify medical research with humans. Specifically, the two philosophical theories that are especially pertinent to acquiring informed consent and upholding autonomy in medical research with humans are deontology and utilitarianism. Although well regarded in philosophical circles, these theories demonstrate pathways for understanding how a clinical ethics consultation curriculum may form. The following section attempts to explain these philosophical theories and demonstrate their role as philosophically based justifications to acquire informed consent and means of upholding human autonomy.

Deontological ethics includes a theory of duty that attempts to reconcile the legitimacy of relationships through moral judgments that become justified by obligatory maxims [111]. These maxims attempt to ground morality in reason by addressing the nature of human beings. Kant claims that human beings possess rationality and are motivated to act morally and work emphatically [112]. Adhering to moral maxims promotes one's flourishing via decision-making capacities, and is [113] In this respect, individuals only act autonomously if their decisions are per one's obligations or maxims. For example, William Frankena's set of theories that attempt to uphold nonmaleficence and beneficence are maxims that attempt to respond to the duty to refrain from committing harm. In this respect, the consent of a research participant is a declaration that requires autonomous choices that can only be legitimate if those choices promote moral obligations that the consenting agent develops for himself. However, the maxims that an individual may establish must withstand the categorical imperative or a standard that determines if maxims are consistent and objective [114].

The categorical imperative attempts to provide uniformity and consistency throughout moral decision-making [115]. The Kantian maxim often associated with bioethics—and is categorically justified—is the categorical imperative that human beings do not use one another as a means for ends butasan end in themselves [116]. This maxim is especially pertinent to medical research with humans and the acquisition of consent when performing research. Although human beings possess autonomy and can make decisions based on their own moral beliefs, they must abide by certain moral obligations under a Kantian framework. For instance, critics may view human subjects who volunteer to test new drugs serve as means to other's ends. However, the same individuals who voluntarily agree to partake in a study choose how they control and conduct their lives [117]. The research participants in this instance may live by a moral obligation or duty that justifies their decision to partake in the study. Kantian deontology allows autonomous choices to take place through a justifiable framework in which moral choices manifest. Rather than possessing autonomy of the will in decision-making, Kant's theory emphasizes the importance of obligatory decision-making that is both following one's moral obligations and discourages emotive influences [118].

Despite the specificity and ample justification behind deontological decision-making, many problems accompany this theory, especially regarding ethical decision-making in medical research. Although categorical imperatives are in place to check and reinforce moral maxims, conflicting obligations become problematic when subscribing to the rigidity of moral maxims [119]. To explore this concern, consider if a researcher promises to provide therapeutic benefits to the blood flow in a research participant's leg. However, the research participant loses his leg in an unrelated incident—the maxim and moral duty the researcher set out to perform tears asunder and impossible to complete.

Thus, the researcher violates his moral maxim despite a conflicting circumstance that was out of his control. Another problematic issue with deontological ethics in research includes the disregard for decision-making based on sympathy and emotion. Kant's deontological theory claims that decisions based on sympathy and emotion have no moral value [120]. However, this moral maxim significantly affects the researcher-participant relationship during a study. Despite a researcher's ability to perform their scientific duties well, the research participant would feel a lack of concern and care on the researcher's part [121]. In this instance, the stringent nature of deontological decision-making affects the necessary compassion in research with humans [122].

Kantian deontology provides an excellent framework for moral decision-making but remains entirely too stringent when addressing a field like medical research with human participants. The variables in medical research with humans are far too vast for an establishment of uniform ethical maxims. However, deontological decision-making in medical research does aid in upholding the autonomy of individuals by providing an additional motive for self-care, namely, a philosophical framework for promoting one's self-respect, value, and motivation [123]. An alternative philosophical theory, which may aid in justifying the ethical framework in which autonomy may uphold ethical research, is utilitarianism.

## Utilitarianism

Utilitarianism is one of the most prominent consequentialist theories. As such, utilitarianism is a philosophical theory that bases its ethical decision-making on value [124]. Contrary to deontology, utilitarianism is not a theory that acts out of duty or a set of maxims. Instead, utilitarianism concentrates on the value of well-being, determined in terms of pleasure, happiness, welfare, and satisfaction [125]. Utilitarianism attempts to make moral decisions in a way that overall good maximizes [126].

Utilitarianism focuses on utility and pleasure rather than the deontological focus of obligatory duty. Utilitarianism asserts that maximum benefit for most agents over harm or detriment yields an ethically beneficial outcome [127]. While this theory certainly is appealing, utilitarianism is unclear on what constitutes maximum good. In theory, if moral agents identify and maximize pleasure or good, human beings could choose options that would yield tremendous benefits. However, choosing an option that yields the best remains problematic because the same decision may not exercise the autonomous rights of individuals, especially in medical research. For example, a research team performing medical research on an experiential drug may require human test subjects to test the drug's efficacy on human physiology. To perform this study ethically, the research team must obtain consent from the research participants. Although researchers ensure the research participants consent to participate amidst the study's risks, their participation will yield a much greater good for a more significant number of people [128].

Many issues arise under the two philosophical theories discussed above. First, following deontological thinking, the human subjects are being treated as a means to an end, but their right to choose stems from a duty that allows them to retain control over their lives [129]. Alternatively, a utilitarian framework asserts that human subjects in the study electively choose to serve as a means to a greater end [130]. Research with human beings is an onerous topic when attempting to uncover the greater good. A balance of benefits and risks requires analysis among a conglomerate of decision-makers, including ethicists, physicians, and research liaisons. Despite the decisions of these individuals, autonomy is a right that is left up to research participants. In this respect, informed consent under a utilitarian framework may manifest as "coerced consent," emphasizing the importance of a research participant's involvement and the greater good his involvement will provide [131].

Deontology and utilitarianism serve as philosophical theories that attempt to justify moral decision-making. Applying these theories to medical research with human beings, a moral agent often uses these theories as justificatory links between autonomous decision-making and informed consent. However, the analysis above demonstrates the complications that may arise when implementing these philosophical theories into acquiring informed consent. In this respect, the debate forces the ethicist to consider what kinds of ethical analyses a professional should use. In the same respect, both deontology and utilitarianism provide a practical standpoint that upholds autonomy and aid ethical decision-making by emphasizing the need for justification in autonomous decisions. Deontology aids ethical decision-making by emphasizing the necessity of a rational will that decides according to values and respect, whereas utilitarianism promotes the maximization of benefits in all ethical decision-making [132].

Despite their attractiveness, deontology and utilitarianism have shortcomings in reconciling proper autonomous decision-making and the ethical acquisition of informed consent in medical research. Specifically, these theories encounter difficulties when attempting to reconcile ethical relativism and vulnerable populations in research. To evaluate a uniform ethical theory for medical research, discussing the issues that arise from applying these theories must be analyzed. In doing so, the benefits of these ethical theories may be examined and applied to an alternative ethical theory.

Due to the importance and demand for human medical research, various theorists have questioned the universal applicability of ethical principles [133]. Although ethical principles like nonmaleficence and beneficence have been established and generally accepted in the research community, various debates emerge regarding the optimization and medium these principles may uphold [134]. Deontology and utilitarianism are commonly applied philosophical mediums, yet various issues that arise with principlistic ethics. There exist many issues that span far beyond the scope of this paper. This analysis examines relativistic issues and the exploitation of vulnerable populations as matters implication by deontology and utilitarianism.

Since deontology focuses on maintaining maxims and adhering to duty, deontological ethics generally focuses on a method that justifies its practice based on consistency. In this respect, a maxim maintains an ethical standard that, when followed, ensures proper ethical practice. However, issues of relativism emerge when formulating deontological maxims. Agreeing upon an excellent deontological maxim becomes especially difficult when developing a uniform protocol for informed consent in medical research [135]. Specific areas in research where informed consent has become difficult include randomization, placebo use, and double-blinding, all of which pertain specifically to research participants [136].

In contemporary research ethics, maxims manifest in the consent documents given to participants [137]. Although these documents cover the legality of research procedures, disclosure and even reflect advancements resulting from the Nuremberg Code, these documents are still not uniform due to the variance in study procedures. Furthermore, these documents sometimes arrive in oral form, which lacks the tangibility of written consent [138]. Even though written consent provides a physical copy of the consent, the concepts and maxims that researchers outline may not be appropriately delineated. In deontology, the maxims that a group of researchers set out to achieve merit if the research participants are aware of these maxims and agree to the duty the researchers attempt to fulfill.

Although proper disclosure of the information is necessary when conducting medical research with human beings, they do not reduce the risks of a study [139]. Nonetheless, disclosing information does ensure that each research participant is responsible for himself [140]. Things are problematic when the maxims formulated by a group of researchers do not coincide with the understanding and safety of the research participants. For instance, researchers may not disclose the results of a study before the results are finalized and published. This scenario is especially disconcerting if the study is therapeutic and sure research participants require the study's health or benefit results. The above instance exemplifies the difficulties behind deontological ethics. However, in fulfilling their duties, another group suffers to an extent. In this respect, deontological ethics cannot serve as the primary vessel that houses and promotes normative ethical principles.

Another issue in medical ethics affected by deontology involves recognizing and respecting a patient or research participant's advanced directive (AD). Typically, AD's are far more common in clinical practice with patients. However, an advanced directive could undoubtedly become a relevant document during medical research. It is not impossible to imagine a research situation where a patient's AD becomes effective due to a lapse in judgment by a research team or a mistake during a drug trial. In these instances, the maxims specified at the beginning of a study would become subservient to the research participant's AD. While researchers are engaging in a deontological structure when obtaining consent from research participants, patients who possess an AD also have a set of maxims that they wish others to refer to during instances of incapacity. While the researcher's maxims are grounded in duties toward knowledge, the research participant's AD maxims are a relational structure toward a duty to oneself and the human person in general [141]. The power that the maxims described in an AD would have over the maxims of researchers illustrates the importance of human duty and demonstrates the difficulties of squaring deontology with autonomy. Philosophers like Emmanuel Levinas, Martin Buber, and Martien Pijnenburg claim that the tremendous emphasis placed on autonomy in contemporary medicine and research is highly problematic due to the burden of decision-making on the research participant or patient. This individual does not possess ample knowledge of the clinical or research protocols [142].

Although the maxims involved in deontology are essential aspects of maintaining ethical duties, abiding by their stringent requirements requires rigidity and a lack of plasticity in the varying world of medical ethics. Medical practice and research work in real-time situational encounters with human beings [143]. By implementing the views and professional opinions of researchers, patients, and research participants, decision-making and the acquisition of consent acquisition can conduct in a compassionate and caring manner [144].

Developing a philosophical theory that optimizes ethical principles and norms is a difficult task. Specifically, forming a philosophical theory becomes especially burdensome when attempting to uphold ethical norms and avoid relativistic notions of its interpretation. The previous section demonstrates the difficulties that accompany deontological ethics and maxim development. While deontology possesses benefits, the theory becomes muddied when one attempts to develop universal maxims that apply to all medical situations. Additionally, the acquisition of informed consent in research remains difficult under deontological ethics due to the influence of researchers and the lack of information for research participants [145]. Although deontological ethics has both benefits and disadvantages, this discussion now turns toward utilitarianism as the dominant philosophical theory in research ethics. Specifically, this section will discuss the effect utilitarian ethics has on vulnerable populations in research.

Although the definition of a vulnerable population or individual may vary, those who classify as such rely on criteria that render them susceptible to exploitation and thus unable to provide consent for a study or procedure [146]. Vulnerable populations have become a focal point of ethical discussion in bioethics [147]. One of the most pressing issues surrounding vulnerable populations involves the inability to grant consent due to capacity, economical, cultural, or cognitive issues [148]. In this respect, the autonomy of individuals is severely affected, not because these individuals are unable to make decisions for themselves, but because these individuals are unable to protect their interests and health [149]. These populations become increasingly enticing to unscrupulous researchers who aim only to obtain results because vulnerable individuals may easily manipulate to gain their consent, albeit unwarranted consent. In this respect, utilitarianism's shortcomings as a philosophical theory prove pretty detrimental. Researchers who intend to gain results from a study to aid a more significant number of people may justify compromising a vulnerable population on utilitarian grounds. For example, individuals who are unable to give consent due to capacity issues may be targeted by researchers because conducting experimental procedures on these individuals would likely be rejected by competent individuals. However, since these individuals are vulnerable, they are more willing to consent due to misunderstanding and miscommunication. Researchers may justify these exploitations because conducting this research will result in a more significant benefit for a larger number of people [150]. Considering this, the ultimate good aims toward a group of individuals, but another group suffers at the expense of knowledge acquisition. In theory, utilitarianism justifies the actions of the researchers. However, a universal philosophical theory must establish principles that protect vulnerable individuals from abuse and exploitation. Unfortunately, utilitarianism leaves various gaps in justification and thus remains problematic.

Although the analysis above demonstrates the strengths of deontology and utilitarianism concerning informed consent in research, various issues arise with these theories. While deontological ethics promotes consistency and aims at dutiful processes, rigorous and stringent criteria leave little room for plasticity and ethical mobility in varying situations. Utilitarianism may provide a morally good outcome for one group but inevitably causes a morally detrimental outcome for another group. To reconcile these theories and to reconcile curriculum points that budding clinical ethicists ought to adhere to, this analysis examines the effectiveness of virtue ethics and the impact it may have on bioethics. While the vastness of research ethics can be overwhelming, the critical component of virtue identification serves as a knowledge and skill aspect of a professional ethicist's training that can aid discussion and virtuous deliberations between involved stakeholders.

## The Morality of Medicine: Theories in Focus

Since the Nuremberg Code, informed consent has aided the ethical practice of medicine and research and promoted the relationship between researchers and research participants. This analysis demonstrates the philosophical justifications of autonomy and autonomy's role in informed consent through deontology and utilitarianism. Nevertheless, to uncover a philosophical basis for medical research with human beings that both upholds autonomy and justifies obtaining consent, the philosophical details of informed consent and autonomy that extend beyond the stipulations of the Nuremberg Code require attention. While the analysis thus far demonstrates the benefits of deontology and utilitarianism, this analysis has also addressed the difficulties accompanying deontology and utilitarianism when attempting to avoid relativistic notions of ethical practice and issues surrounding vulnerable populations. This alternative theory must abide by a standard of professionalism that maintains the benefits of deontology and utilitarianism while simultaneously evacuating the negative aspects of these theories.

The proper application of moral principles and rules maintains the ethical norms necessary for human engagement and promotes moral excellence in medical research. Promoting moral principles specific virtues that abide by moral norms must be cultivated. Moral virtue pertains to character traits that are dispositional and reliable [151]. In respect to medical research with humans, moral virtue must follow by a common morality that, when appropriately executed, exercises a means of engagement with other human beings that respects the rights of others and justifies engagement with other human beings [152]. If proper virtue is cultivated and promoted as an integral part of human medical research, crimes and ethical disservices may be avoided and promote the commonality of moral norms [153].

Virtue and moral norms become enacted when vocational disciplines have a universal understanding of the goals and ends [154]. Despite the specific knowledge healthcare professionals and researchers must relay to their participants, certain social and professional expectations require consensus. Researchers must possess virtues while conducting medical research include compassion, discernment, trustworthiness, integrity, and conscientiousness [155].

If exercised correctly, these five virtues cultivate responsibilities that coincide with a professional code of ethics [156]. The first focal virtue is compassion, which requires the capacity to have a sympathetic understanding of a research participant's current situation. This virtue must be recognized and regulated adequately with the participant's emotions. If the participant's emotions are too heightened, the researcher's understanding and care should not be appropriately understood and tear the relationship between the researcher and the research participant [157]. The second focal virtue is discernment. Discernment pertains to the Aristotelian understanding of practical wisdom (φρόνεσις). Practical wisdom involves calculated and reasonable decision-making in situations that call for rationality [158]. The next virtue, trustworthiness, is essential because it is the foundation of a reliable patient-researcher relationship. Trustworthiness is a virtue that entails confident reliance and dependence on another individual [159]. Next, integrity involves the capacity of an individual to abide by the rules and principles set before them. Finally, conscientiousness grants an individual the moral capacity for decision-making [160]. While special orders or conflicting issues with institutional compliance may arise, the genuinely virtuous individual will exercise discretion in decision-making and focus on making calculated decisions when dealing with moral issues in medical research [161].

The importance of moral virtue in medical practice and medical research with humans is of the utmost importance [162]. Providing a philosophical basis for morality justifies medical research with humans and provides a structure of morals in medicine. However, the moral virtues discussed above are still debated and not viewed as the authoritative and justifiable philosophical foundations for medical professionals and researchers. With the discussion of moral virtues of medicine explained, the next section of this analysis addresses how an ethics curriculum provides a foundational structure for ethics knowledge in medicine [163].

Though there is consensus amongst professionals regarding the foundational moral obligations in medicine, Medical practitioners and researchers do not necessarily agree upon the methods of adhering to their obligations [164]. Albeit unconsciously, medical practitioners often make vocational moral decisions rooted in deontology and utilitarianism. Moral action remains incomplete without consensus and is significantly fractured when dealing with ethical relativism and vulnerable populations. Here, we uncover philosophical theories and their shortcomings. In doing so, a discussion of medical morality outlines the requirements and expectations of medical researchers and provides a basis of morality that a philosophical theory must follow.

Medical ethics inherently questions whether a justified philosophical theory can maintain the moral obligations of health care professionals [165]. However, identifying this theory becomes difficult due to the vast differences in medical situations and medical research. Due to the variance in opinions from medical professionals, theorists, researchers, and philosophers, issues of moral permissibility arise [166]. A relativistic attitude in medicine and medical research inevitably develops when individuals have varying opinions on ethical matters. The relativism that accompanies contemporary biomedical issues calls for a common morality in medicine [167].

To uncover a common morality in medicine and medical research, theorists commonly search for the nature of what makes a medical researcher an excellent medical researcher [168]. A philosophy that fuses the nature of medical research with rational analysis adds disciplined elements and criteria to the process of ethical decision-making in research [169]. This process aids in ethical decision-making on the researcher's part and avoids ambiguity or superfluous practices in ethical decision-making. Most importantly, this process provides evidence of the effectiveness and practicality of the proposed critical components for a professional curriculum in clinical ethics consultations.

To suture the divide between philosophical principles and decisions in medical research, the dialectical amalgam of judgments and morals must guide this union through specific contexts [170]. Partaking in dialectical conflict resolution and implementing virtue may resolve ethical discrepancies [171]. This process involves defining the specifics of a study, clarifying the language of medicine and philosophy, and uncovering relevant virtues. Doing so aids the process of establishing a philosophical framework for ethics in research education.Implementing philosophical practice in medical research reaffirms the commitments of the individuals involved [172]. The common morality serves as a universal ethical grounding that multi-faceted and variant ethical dilemmas may direct their end. Common morality serves as an applied ethic that develops general principles, while the researcher-philosopher serves as a medium in which these principles apply [173]. However, to achieve a common moral standing in research ethics, a philosophical theory capable of adaptation must regulate the varying situation that may develop. Initially inducted by Aristotle, virtue ethics prove a relevant philosophical theory that aids ethical decision-making in research.

Virtue ethics serves as an alternative philosophical medium in which ethical principles may be practiced and upheld. Virtue ethics acts out of acare position and derives from the classical Greek tradition of Aristotelian ethics [174]. Virtue ethics involves ethical decision-making that revolves around the idea of practical virtue, that is, the capacity of performing or functioning well [175]. Although the term has been translated and understood as virtue or excellence, it has lost its primary meaning as a term that embodies both moral and practical virtues [176]. Earlier in this analysis, deontological ethics, utilitarianism, and moral excellence identified the importance of praxis.

Furthermore, this analysis identifies one understanding of practical virtue by outlining the moral virtues of medical research and practice in section five. This analysis now turns toward the practical virtues that reside withinhealth care. In doing so, we identify virtue ethics as a philosophy that aids ethical decision-making when obtaining consent from research participants.

The virtue approach to medical ethics hinges upon the Aristotelian understanding of virtue ethics as a vocational practice. Bioethics' traditional principles are autonomy, beneficence, nonmaleficence, and justice, all of which require virtue to practice successfully [177]. However, in *Nicomachean Ethics*, Aristotle distinguishes explicitly between practical and moral virtues and explains why they relate [178].Aristotle claims that individuals must discern a mean or mid-point understanding of two extremes to achieve practical wisdom. For instance, the virtue of courage is the mean point between rashness and cowardice [179].

If an individual can rationally calculate his circumstances, he may choose a reasonable middle-ground between two extreme outcomes [180]. For this reason, autonomous choices that are informed and reasoned well are of the utmost importance when attempting to identify ethical practice. Although an autonomous individual may choose freely, virtuous decisions must be made from correctness and strive toward happiness as an end or *telos* [181]. In this respect, Kant's deontological approachaligns with Aristotle's virtue ethics due to both philosophies' responsibility toward an end. However, while deontological ethics remains exceptionally rigid in its formulation of maxims and rejects emotive notions of care as motivation for ethical practice, virtue ethics presents various tools or virtues that may require moral resolution. Furthermore, virtue ethics' ability to adjust and adapt to various situations heavily recognizes practical and moral virtues [182].

Virtue ethics is unique because it recognizes and implements practical virtues throughout its application. A practical virtue may be carpentry or sewing. However, for this analysis, medicine and research are especially pertinent. For Aristotle, it is possible for a medical researcher failed to uphold ethical practice by not exercising his practical virtues. Both moral and practical virtues must strive toward a mean [183]. In this respect, Aristotle discerns between practical intelligence and moral virtues. To uphold virtuous practice or exercise practical intelligence, an agent must exercise moral virtues appropriately [184].

This understanding of Aristotelian virtue ethics demonstrates the need for calculated discernment between two extremes of practice and the importance of upholding a middle ground [185].

The framework of virtue ethics may apply as an alternative philosophy to deontological and utilitarian ethics in research. Alasdair MacIntyre uses Aristotle'smoral and practical virtues theoryto solve relativistic notions of ethics by creating sects of ideas among individuals. MacIntyre's book is appropriately named, considering the framework of his discussion. Once ethics has been established and identified, MacIntyre uses virtue ethics' tools to resolve relativistic notions in ethics. MacIntyre's solution may be applied and used as an alternative philosophy to the commonly accepted standards of deontology and utilitarianism in medical research and medical ethics.

This analysis demonstrates the shortcomings of deontology in medical research by objecting to its rigidity when developing maxims and thus causing a lack of ethical uniformity. Furthermore, this analysis demonstrated the shortcomings of utilitarianism by uncovering the danger this philosophy possesses of exploiting individuals in medical research. MacIntyre's virtue ethics resolve these discrepancies by creating sects of ideals that may uphold within a community [186]. Medical researchers may make up a community with the moral virtues in medical research making up the common morality within that community. Practical virtues like an attentive radiologist, an accurate statistician, or a data analyst are vocational virtues subject to change depending on the study [187]. If researchers and research participants aim toward ethical ends by communicating with each other as a community, medical research's common morality may establish under moral and practical virtues. This process then bolsters practical autonomous decisions and justifies the acquisition of consent for researchers and research participants.

This analysis demonstrates that analytic moral reasoning skills are crucial in health care ethics consultation and are highly beneficial to effective facilitation methods. By examining the philosophical Kantian ends of self-improvement and the improvement of others, ethicists understandthe moral duty involved in their practice. Kant's philosophy opens a doorway through which ethicists may pursue key elements in moral analytic reasoning, namely, virtues. If health care ethics consultants grasp the significance of virtue and its relationship to performing functions well and striving toward the Kantian ends discussed in this analysis, facilitation and conflict resolution efforts become more effective. Aiding an ethicist's ability to discern between warranted and unwarranted emotions in consultation settings allows the ethics consultant to critically engage in practical consultation skills and conversation.

By examining the nature of virtue and understanding how to conduct consultations virtuously, health care ethics consultants may engage with their patients, patient's family members, and other health care professionals more effectively and efficiently. The issues these theories present become apparent when considering autonomous decision-making in informed consent in research ethics. Through addressing the origins of formal informed consent via a historical analysis of the Nuremberg Code, deontological and utilitarian ethics serve as theories that attempt to promote informed consent in research that uphold the stipulations of the Nuremberg Code. However, the theories' difficulties present inherent threats to autonomy and autonomous decision-making due to the lack of information for research participants.

The rigidity of deontology and the biases of utilitarianism present several issues when attempting to develop a philosophical, moral theory that clinical ethicists may follow. By outlining the importance of moral virtue in medicine, virtue ethics presents itself as an alternative philosophical theory that bolsters human autonomy by upholding virtues that inevitably result in proper information disclosure and support from medical professionals. Additionally, virtue ethics presents itself as a philosophy that maintains the beneficial relationships of all parties involved in medical research. Addressing medical research through the philosophy of virtue ethics presents the discipline as a community and fosters beneficial partnerships and results. If virtue ethics is implemented properly into a core curriculum for clinical ethics, standards of moral excellence, specifically concerning autonomy and informed consent, are upheld and promote the acquisition of medical knowledge and mutual respect for human beings.

While the discussion in chapter one demonstrates the philosophical groundwork upon which clinical ethicists may base their skillset through consultation methodology, the application and measurement of these skills and knowledge points are inadequate without a proper evaluation method. By exploring organizational venues in which key curriculum components manifest, clinical ethicists become better equipped to professionalize their discipline further. Furthermore, an evaluation of the organizational structures associated with integrating a formalized ethics curriculum bolsters this analysis's goal at implementing key virtue components to an ethics consultation curriculum.

# References

1. Jonsen, Albert R., Mark Siegler, and William J. Winslade. *Clinical Ethics: A Practical Approach to Ethical Decisions in Clinical Medicine*. 8th ed. (New York: McGraw Hill, Medical Pub. Division, 2015), 2–6; *See* Bandura, Albert, Claudio Barbaranelli, Gian Vittorio Caprara, and Concetta Pastorelli, "Mechanisms of moral disengagement in the exercise of moral agency," *Journal of personality and social psychology* 71, no. 2 (1996): 364.
2. Hittinger, Russell. *A Critique of the New Natural Law Theory*, (Notre Dame, IN: University of Notre Dame Press, 1987), 30–42; cf. Kant, Immanuel. *What does it mean to orient oneself in thinking?*. Daniel, Fidel, Ferrer, Verlag. (Cambridge: Cambridge University Press 2014), 3–17.
3. Jonsen, Albert R., Mark Siegler, and William J. Winslade. *Clinical Ethics: A Practical Approach to Ethical Decisions in Clinical Medicine*. 8th ed. (New York: McGraw Hill, Medical Pub. Division, 2015), 3–10. *See* Gillon, Raanan. "Medical ethics: four principles plus attention to scope." *BMJ: British Medical Journal* 309, no. 6948 (1994): 184.
4. Baylis, Françoise, "Health Care Ethics Consultation: 'Training in Virtue'," In *Performance, Talk, Reflection*. (Springer Netherlands, 1999), 25–41.
5. Louden, Robert B. "Kant's Virtue Ethics." *Philosophy* 61, no. 238 (1986): 473–489. *See* Garbutt, Gerard, and Peter Davies. "Should the practice of medicine be a deontological or utilitarian enterprise?," *Journal of medical ethics* 37, no. 26, (2011): 267–270.
6. Louden, Robert B. "Kant's Virtue Ethics." *Philosophy* 61, no. 238 (1986): 473–474. *See* McCarthy, Gerald. "A Via Media Between Skepticism and Dogmatism?: Newman's and MacIntyre's Anti-Foundationalist Strategies." *Newman Studies Journal* 6, no. 2 (2009): 57–81; cf. Foot, Philippa. "Morality as a system of hypothetical imperatives," *The Philosophical Review* 81, no. 3 (1972): 305–316; cf. Hursthouse, Rosalind, Gavin Lawrence, and Warren Quinn. *Virtues and reasons: Philippa Foot and moral theory: essays in honour of Philippa Foot*, (Oxford: Oxford University Press 1995), 13–23.
7. Louden, Robert B. "Kant's Virtue Ethics." *Philosophy* 61, no. 238 (1986): 475–476 *See* Nussbaum, Martha C. "Virtue ethics: a misleading category?." *The Journal of Ethics* 3, no. 3 (1999): 163–201; cf Devettere, Raymond J. *Introduction to virtue ethics: Insights of the ancient Greeks*, (Georgetown University Press, 2002), 22–24.
8. Honig, Bonnie. *Political theory and the displacement of politics*, (Ithaca: Cornell University Press, 1993), 18–20; *See* Heubel, Friedrich, and Nikola Biller-Andorno, "The contribution of Kantian moral theory to contemporary medical ethics: a critical analysis," *Medicine, Health Care and Philosophy* 8, no. 1 (2005): 5–18.
9. Louden, Robert B. "Kant's Virtue Ethics." *Philosophy* 61, no. 238 (1986): 477–478. *See* Wright, Thomas A., and Jerry Goodstein. "Character is not "dead" in management research: A review of individual character and organizational-level virtue." *Journal of Management* 33, no. 6 (2007): 928–958.
10. Louden, Robert B. "Kant's Virtue Ethics." *Philosophy* 61, no. 238 (1986): 482–488; *See* Dierksmeier, Claus, "Kant on virtue," *Journal of Business Ethics* 113, no. 4 (2013): 597–609.
11. Honig, Bonnie. *Political theory and the displacement of politics*, (Ithaca: Cornell University Press, 1993), 20–24; *See* Wright, R. George, "Treating Persons as Ends in Themselves: The Legal Implications of a Kantian Principle." *U. Rich. l. Rev.* 36 (2002): 271;cf. Curzer, Howard J. "Aristotle: Founder of the ethics of care," *The Journal of Value Inquiry* 41, no. 2–4 (2007): 221–243.
12. Louden, Robert B. "Kant's Virtue Ethics." *Philosophy* 61, no. 238 (1986): 484–488.
13. Baylis, Françoise. "Health Care Ethics Consultation: 'Training in Virtue'." *Performance, Talk, Reflection*, 1999, 25–27; *See* Cohen, Eric, "Conservative Bioethics & the Search for Wisdom," *Hastings Center Report* 36, no. 1 (2006): 44–56.
14. Baylis, Françoise, "Health Care Ethics Consultation: 'Training in Virtue'," In *Performance, Talk, Reflection*. (Springer Netherlands, 1999), 28–30; *See* Baylis, Françoise. "Health Care

Ethics Consultation: 'Training in Virtue'." In *Performance, Talk, Reflection*, (Springer Netherlands, 1999), 25–41.

15. Johnson, Robert N. "Kant's Conception of Virtue." *Annual Review of Law and Ethics* 5 (1997). 365–367.

16. Baylis, Françoise, "Health Care Ethics Consultation: 'Training in Virtue'," In *Performance, Talk, Reflection*, (Springer Netherlands, 1999), 30–32; *See* Kish-Gephart, Jennifer J., David A. Harrison, and Linda Klebe Treviño. "Bad apples, bad cases, and bad barrels: meta-analytic evidence about sources of unethical decisions at work," *The Journal of Applied Psychology* 95, no. 4, (2010): 791.

17. Baylis, Françoise, "Health Care Ethics Consultation: 'Training in Virtue'," In *Performance, Talk, Reflection*, (Springer Netherlands, 1999), 33–34.

18. Baylis, Françoise, "Health Care Ethics Consultation: 'Training in Virtue'," In *Performance, Talk, Reflection*, (Springer Netherlands, 1999), 35–36.

19. Gao, Guoxi. "Kant's virtue theory." *Frontiers of Philosophy in China* 5, no. 2 (2010): 271–278. *See* Micciche, Laura. *A way to move: Rhetorics of emotion and composition studies*, (Boynton/Cook Heinemann, 2003) 2–4.

20. Baylis, Françoise, "Health Care Ethics Consultation: 'Training in Virtue'," In *Performance, Talk, Reflection*, (Springer Netherlands, 1999), 36–40; *See* La Puma, John M. D., and David L. Schiedermayer. "Ethics consultation: skills, roles, and training." *Annals of Internal Medicine* 114 (1991): 155–160.

21. Johnson, Robert N. "Kant's Conception of Virtue." *Annual Review of Law and Ethics* 5 (1997). 370–387.

22. Dierksmeier, Claus. "Kant on Virtue." *Journal of Business Ethics* 113, no. 4 (2013): 597–609.

23. Korsgaard, Christine M. "Aristotle on Function and Virtue." *The Constitution of Agency*, 2008. 259–262.

24. Honig, Bonnie. *Political theory and the displacement of politics*. Ithaca: Cornell University Press, 1993. 20–25.

25. Roberts, Robert C. "Aristotle on virtues and emotions." *Philosophical Studies* 56, no. 3 (1989). 293–296.

26. Gotthelf, Allan, James G. Lennox, and Lester H. Hunt. *Metaethics, egoism, and virtue: studies in Ayn Rand's normative theory*. Pittsburgh, PA: University of Pittsburgh Press, 2011. 149–157.

27. Roberts, Robert C. "Aristotle on virtues and emotions." *Philosophical Studies* 56, no. 3 (1989). 293–296.

28. Førtenbaugh, William W., *Aristotle on emotion: a contribution to philosophical psychology, rhetoric, poetics, politics, and ethics*, (Duckworth 2002), 20–34.

29. Aristotle, J. A. K. Thomson, and Hugh Tredennick. *The ethics of Aristotle: the Nicomachean ethics*. Harmondsworth: Penguin, 2004. 31–49; 1103a–1109b.

30. Arnold, Robert M., MD, Kenneth A. Berkowitz, MD, Nancy Neveloff Dubler, LLB, Denise Dudzinski, PhD MTS, Ellen Fox, MD, Andrea Frolic, PhD, Jacqueline J. Glover, PhD, Kenneth Kipnis, PhD, Ann Marie Natali, MBA, William A. Nelson, PhD, Mary V. Rorty, PhD, Paul M. Schyve, MD, Joy D. Skeel, MDiv, and Anita J. Tarzian, PhD RN. *Core competencies for healthcare ethics consultation: the report of the American Society for Bioethics and Humanities*. 2nd Ed. (Glenview, IL: ASBH, American Society for Bioethics and Humanities, 2011), 19–32.

31. Roberts, Robert C. "Aristotle on virtues and emotions." *Philosophical Studies* 56, no. 3 (1989). 296–297.

32. Roberts, Robert C. "Aristotle on virtues and emotions." *Philosophical Studies* 56, no. 3 (1989). 296–297. *See* Fortenbaugh, William W. "Aristotle on emotion: a contribution to philosophical psychology, rhetoric, poetics, politics, and ethics." (2002).

33. Roberts, Robert C. "Aristotle on virtues and emotions." *Philosophical Studies* 56, no. 3 (1989): 297–299; *See* Sokolon, Marlene Karen. *Political emotions: Aristotle and the symphony of reason and emotion*, (Northern Illinois University: Ullinois University Press 2003), 1–23.

34. Plato. "Laches" Plato Complete Works. Ed. Edith Hamilton and Huntington Cairns. New York: Pantheon, 1961. N. Print; *See*Irwin, Terence. *Plato's ethics*, (Oxford University Press, 1995), 2–3.
35. Roberts, Robert C. "Aristotle on virtues and emotions." *Philosophical Studies* 56, no. 3 (1989). 298–305.
36. Halwani, Raja. "Care Ethics and Virtue Ethics." *Hypatia* 18, no. 3 (2003). 161–169.
37. Halwani, Raja. "Care Ethics and Virtue Ethics." *Hypatia* 18, no. 3 (2003). 170. Cf. Krantz, David H., and Howard C. Kunreuther. "Goals and plans in decision-making." *Judgment and Decision-making* 2, no. 3 (2007): 137.
38. Dahl, Norman O. *Practical reason, Aristotle, and weakness of the will*. Minneapolis: Minn., 1984. 123–127. *See* Lacour-Gayet, François, David R. Clarke, and Aristotle Committee. "The Aristotle method: a new concept to evaluate quality of care based on complexity." *Current opinion in pediatrics* 17, no. 3 (2005): 412–417.
39. Halwani, Raja. "Care Ethics and Virtue Ethics." *Hypatia* 18, no. 3 (2003). 170–175. *See* Schwartz, Barry, and Kenneth E. Sharpe. "Practical wisdom: Aristotle meets positive psychology." *Journal of Happiness Studies* 7, no. 3 (2006): 377–395.
40. Oliver, Paul. "Conclusion: The Role of The Researcher." In *The Student's Guide to Research Ethics*. 2nd ed. Maidenhead, (Berkshire, England: McGraw-Hill/Open University Press, 2010), 172. *See* Fletcher, Joseph F., *Morals and Medicine: the moral problems of the patient's right to know the truth, contraception, artificial insemination, sterilization, euthanasia*, (Princeton University Press, 2015), 4–40.
41. Oliver, Paul. "Conclusion: The Role of The Researcher." In *The Student's Guide to Research Ethics*. 2nd ed. Maidenhead, (Berkshire, England: McGraw-Hill/Open University Press, 2010), 172.
42. Baylis, Françoise, and Francoise Baylis. "Heroes in Bioethics." *The Hastings Center Report* 30, no. 3 (2000). 37–38.
43. Maitlis, Sally, and Hakan Ozcelik. "Toxic Decision Processes: A Study of Emotion and Organizational Decision-making." *Organization Science* 15, no. 4 (2004). 381–391. *See* Begley, Ann M. "Facilitating the development of moral insight in practice: teaching ethics and teaching virtue." *Nursing Philosophy* 7, no. 4 (2006): 257–265.
44. Dahl, Norman O. *Practical reason, Aristotle, and weakness of the will*. Minneapolis: Minn., 1984. 119–123.
45. Pellegrino, Edmund D., H. Tristram Engelhardt, and Fabrice Jotterand. *The philosophy of medicine reborn: a Pellegrino reader*. Notre Dame: University of Notre Dame Press, 2011. 147–159; *See* Mackenzie, Catriona, Wendy Rogers, and Susan Dodds, eds. *Vulnerability: New essays in ethics and feminist philosophy*. (Oxford University Press, 2014), 13–52.
46. Pellegrino, Edmund D., H. Tristram Engelhardt, and Fabrice Jotterand. *The philosophy of medicine reborn: a Pellegrino reader*. Notre Dame: University of Notre Dame Press, 2011. 50–54.
47. Kraut, Richard. "Aristotle's Ethics." Stanford Encyclopedia of Phlosophy. Stanford University, June 15, 2018. https://plato.stanford.edu/entries/aristotle-ethics/#DoctMean.
48. Arnold, Robert M., MD, Stuart J. Youngner, MD, Mark P. Aulisio, PhD, Françoise Baylis, PhD, Charles Bosk, PhD, Dan Brock, PhD, Howard Brody, MD PhD, Linda Emanuel, MD PhD, Arlene Fink, PhD, John Fletcher, PhD, Jacqueline J. Glover, PhD, George Kanoti, STD, Steven Miles, MD, Kathryn Moseley, MD, William Nelson, PhD, Ruth Purtilo, PhD, Cindy Rushton, DNSc MSN, Paul Schyve, MD, Melanie H. Wilson Silver, MA, Joy Skeel, MDiv BSN, and William Winslade, PhD JD. *Core competencies for healthcare ethics consultation: the report of the American Society for Bioethics and Humanities*. (Glenview, IL: American Society for Bioethics and Humanities, 1998), 1–2. cf; Tarzian, Anita J., and ASBH Core Competencies Update Task Force, "Health care ethics consultation: An update on core competencies and emerging standards from the American Society for Bioethics and Humanities' Core Competencies Update Task Force," *The American Journal of Bioethics* 13, no. 2 (2013): 3–13.

49. Arnold, Robert M., MD, Stuart J. Youngner, MD, Mark P. Aulisio, PhD, Françoise Baylis, PhD, Charles Bosk, PhD, Dan Brock, PhD, Howard Brody, MD PhD, Linda Emanuel, MD PhD, Arlene Fink, PhD, John Fletcher, PhD, Jacqueline J. Glover, PhD, George Kanoti, STD, Steven Miles, MD, Kathryn Moseley, MD, William Nelson, PhD, Ruth Purtilo, PhD, Cindy Rushton, DNSc MSN, Paul Schyve, MD, Melanie H. Wilson Silver, MA, Joy Skeel, MDiv BSN, and William Winslade, PhD JD. *Core competencies for healthcare ethics consultation: the report of the American Society for Bioethics and Humanities.* (Glenview, IL: American Society for Bioethics and Humanities, 1998), 1–2.

50. Annas, George J. "Will the Real Bioethics (Commission) Please Stand up?" *The Hastings Center Report* 24, no. 1 (1994). 19–21. *See* Churchill, Larry R. "Are we professionals? A critical look at the social role of bioethicists," *Daedalus* 128, no. 4 (1999): 253–274.

51. Carmichael, Peter A. "For Want of Reason and Ethics." *The Journal of Philosophy* 44, no. 3 (1947). 67–79.

52. Dougherty, Charles J. *Ideal, fact, and medicine: a philosophy for health care.* Lanham u.a.: Univ. Pr. of America, 1985. 94–99.

53. Arnold, Robert M., MD, Kenneth A. Berkowitz, MD, Nancy Neveloff Dubler, LLB, Denise Dudzinski, PhD MTS, Ellen Fox, MD, Andrea Frolic, PhD, Jacqueline J. Glover, PhD, Kenneth Kipnis, PhD, Ann Marie Natali, MBA, William A. Nelson, PhD, Mary V. Rorty, PhD, Paul M. Schyve, MD, Joy D. Skeel, MDiv, and Anita J. Tarzian, PhD RN. *Core competencies for healthcare ethics consultation: the report of the American Society for Bioethics and Humanities.* 2nd Ed. Glenview, IL: ASBH, American Society for Bioethics and Humanities, 2011. 32.

54. Arnold, Robert M., MD, Kenneth A. Berkowitz, MD, Nancy Neveloff Dubler, LLB, Denise Dudzinski, PhD MTS, Ellen Fox, MD, Andrea Frolic, PhD, Jacqueline J. Glover, PhD, Kenneth Kipnis, PhD, Ann Marie Natali, MBA, William A. Nelson, PhD, Mary V. Rorty, PhD, Paul M. Schyve, MD, Joy D. Skeel, MDiv, and Anita J. Tarzian, PhD RN. *Core competencies for healthcare ethics consultation: the report of the American Society for Bioethics and Humanities.* 2nd Ed. Glenview, IL: ASBH, American Society for Bioethics and Humanities, 2011. 26.

55. Arnold, Robert M., MD, Stuart J. Youngner, MD, Mark P. Aulisio, PhD, Françoise Baylis, PhD, Charles Bosk, PhD, Dan Brock, PhD, Howard Brody, MD PhD, Linda Emanuel, MD PhD, Arlene Fink, PhD, John Fletcher, PhD, Jacqueline J. Glover, PhD, George Kanoti, STD, Steven Miles, MD, Kathryn Moseley, MD, William Nelson, PhD, Ruth Purtilo, PhD, Cindy Rushton, DNSc MSN, Paul Schyve, MD, Melanie H. Wilson Silver, MA, Joy Skeel, MDiv BSN, and William Winslade, PhD JD. *Core competencies for healthcare ethics consultation: the report of the American Society for Bioethics and Humanities.* (Glenview, IL: American Society for Bioethics and Humanities, 1998), 21. *See* King, Nancy MP., "Who ate the apple? A commentary on the core competencies report." In *HEC Forum*, vol. 11, no. 2, Springer Netherlands, (1999): 170–175.

56. Chidwick, Paula, Karen Faith, Dianne Godkin, and Laurie Hardingham. "Clinical education of ethicists: the role of a clinical ethics fellowship." *BMC Medical Ethics* 5, no. 1 (2004). 5–6.

57. Chidwick, Paula, Karen Faith, Dianne Godkin, and Laurie Hardingham. "Clinical education of ethicists: the role of a clinical ethics fellowship." *BMC Medical Ethics* 5, no. 1 (2004). 4–8.

58. Kultgen, John. *Ethics and Professionalism*, (Philadelphia: University of Pennsylvania Press, 2010): 257–262.

59. Arnold, Robert M., MD, Stuart J. Youngner, MD, Mark P. Aulisio, PhD, Françoise Baylis, PhD, Charles Bosk, PhD, Dan Brock, PhD, Howard Brody, MD PhD, Linda Emanuel, MD PhD, Arlene Fink, PhD, John Fletcher, PhD, Jacqueline J. Glover, PhD, George Kanoti, STD, Steven Miles, MD, Kathryn Moseley, MD, William Nelson, PhD, Ruth Purtilo, PhD, Cindy Rushton, DNSc MSN, Paul Schyve, MD, Melanie H. Wilson Silver, MA, Joy Skeel, MDiv BSN, and William Winslade, PhD JD. *Core competencies for healthcare ethics consultation: the report of the American Society for Bioethics and Humanities.* (Glenview, IL: American Society for Bioethics and Humanities, 1998), 24–26.

60. Kultgen, John. *Ethics and Professionalism*, (Philadelphia: University of Pennsylvania Press, 2010): 269–273.
61. Kultgen, John. *Ethics and Professionalism*, (Philadelphia: University of Pennsylvania Press, 2010): 269–272.
62. Reiter-Theil, S. "The Freiburg approach to ethics consultation: process, outcome and competencies." *Journal of Medical Ethics* 27, no. 90001 (2001). 21–23.
63. Kultgen, John. *Ethics and Professionalism*, (Philadelphia: University of Pennsylvania Press, 2010): 264–269.
64. Kodish, Eric, Joseph J. Fins, Clarence Braddock, Felicia Cohn, Nancy Neveloff Dubler, Marion Danis, Arthur R. Derse, Robert A. Pearlman, Martin Smith, Anita Tarzian, Stuart Youngner, and Mark G. Kuczewski. "Quality Attestation for Clinical Ethics Consultants: A Two-Step Model from the American Society for Bioethics and Humanities." *Hastings Center Report* 43, no. 5 (2013): 27–30.
65. Bayley, Carol. "The Next Step for Quality Attestation." *Hastings Center Report* 43, no. 5 (2013). 37–39.
66. Pierce, Jessica, and George Randels. *Contemporary bioethics: a reader with cases*. New York: Oxford University Press, 2010. 684–689.
67. Gaudine, Alice, and Linda Thorne. "Emotion and Ethical Decision-Making in Organizations." *Journal of Business Ethics* 31, no. 2 (May 2001). 176–177.
68. Gaudine, Alice, and Linda Thorne. "Emotion and Ethical Decision-Making in Organizations." *Journal of Business Ethics* 31, no. 2 (May 2001). 177–181.
69. Gaudine, Alice, and Linda Thorne. "Emotion and Ethical Decision-Making in Organizations." *Journal of Business Ethics* 31, no. 2 (May 2001). 183–184.
70. Gaudine, Alice, and Linda Thorne. "Emotion and Ethical Decision-Making in Organizations." *Journal of Business Ethics* 31, no. 2 (May 2001). 174–185.
71. Micciche, Laura. "Emotion, Ethics, and Rhetorical Action." *JAC* 25, no. 1 (2005): 161–164.
72. Arnold, Robert M., MD, Kenneth A. Berkowitz, MD, Nancy Neveloff Dubler, LLB, Denise Dudzinski, PhD MTS, Ellen Fox, MD, Andrea Frolic, PhD, Jacqueline J. Glover, PhD, Kenneth Kipnis, PhD, Ann Marie Natali, MBA, William A. Nelson, PhD, Mary V. Rorty, PhD, Paul M. Schyve, MD, Joy D. Skeel, MDiv, and Anita J. Tarzian, PhD RN. *Core competencies for healthcare ethics consultation: the report of the American Society for Bioethics and Humanities*. 2nd Ed. (Glenview, IL: ASBH, American Society for Bioethics and Humanities, 2011), 19–21.
73. Devettere, Raymond J. *Practical decision-making in health care ethics: cases, concepts, and the virtue of prudence*. Washington, D.C.: (Georgetown University Press, 2016), 346–352.
74. Arnold, Robert M., MD, Kenneth A. Berkowitz, MD, Nancy Neveloff Dubler, LLB, Denise Dudzinski, PhD MTS, Ellen Fox, MD, Andrea Frolic, PhD, Jacqueline J. Glover, PhD, Kenneth Kipnis, PhD, Ann Marie Natali, MBA, William A. Nelson, PhD, Mary V. Rorty, PhD, Paul M. Schyve, MD, Joy D. Skeel, MDiv, and Anita J. Tarzian, PhD RN. *Core competencies for healthcare ethics consultation: the report of the American Society for Bioethics and Humanities*. 2nd Ed. (Glenview, IL: ASBH, American Society for Bioethics and Humanities, 2011), 19–21.
75. Tollefsen, Christopher. "Practical Reason and Ethics above the Line." *Ethical Theory and Moral Practice* 5, no. 1 (March 2002). 67–70.
76. Arnold, Robert M., MD, Kenneth A. Berkowitz, MD, Nancy Neveloff Dubler, LLB, Denise Dudzinski, PhD MTS, Ellen Fox, MD, Andrea Frolic, PhD, Jacqueline J. Glover, PhD, Kenneth Kipnis, PhD, Ann Marie Natali, MBA, William A. Nelson, PhD, Mary V. Rorty, PhD, Paul M. Schyve, MD, Joy D. Skeel, MDiv, and Anita J. Tarzian, PhD RN. *Core competencies for healthcare ethics consultation: the report of the American Society for Bioethics and Humanities*. 2nd Ed. (Glenview, IL: ASBH, American Society for Bioethics and Humanities, 2011), 22–23.
77. Arnold, Robert M., MD, Kenneth A. Berkowitz, MD, Nancy Neveloff Dubler, LLB, Denise Dudzinski, PhD MTS, Ellen Fox, MD, Andrea Frolic, PhD, Jacqueline J. Glover, PhD,

Kenneth Kipnis, PhD, Ann Marie Natali, MBA, William A. Nelson, PhD, Mary V. Rorty, PhD, Paul M. Schyve, MD, Joy D. Skeel, MDiv, and Anita J. Tarzian, PhD RN. *Core competencies for healthcare ethics consultation: the report of the American Society for Bioethics and Humanities*. 2ⁿᵈ Ed. (Glenview, IL: ASBH, American Society for Bioethics and Humanities, 2011), 23.

78. Dunn, Barnaby D., Hannah C. Galton, Ruth Morgan, Davy Evans, Clare Oliver, Marcel Meyer, Rhodri Cusack, Andrew D. Lawrence, and Tim Dalgleish. "Listening to Your Heart." *Psychological Science* 21, no. 12 (2010). 1842–1843.

79. Arnold, Robert M., MD, Kenneth A. Berkowitz, MD, Nancy Neveloff Dubler, LLB, Denise Dudzinski, PhD MTS, Ellen Fox, MD, Andrea Frolic, PhD, Jacqueline J. Glover, PhD, Kenneth Kipnis, PhD, Ann Marie Natali, MBA, William A. Nelson, PhD, Mary V. Rorty, PhD, Paul M. Schyve, MD, Joy D. Skeel, MDiv, and Anita J. Tarzian, PhD RN. *Core competencies for healthcare ethics consultation: the report of the American Society for Bioethics and Humanities*. 2ⁿᵈ Ed. (Glenview, IL: ASBH, American Society for Bioethics and Humanities, 2011), 24.

80. Maitlis, Sally, and Hakan Ozcelik. "Toxic Decision Processes: A Study of Emotion and Organizational Decision-making." *Organization Science* 15, no. 4 (2004). 376–377.

81. Maitlis, Sally, and Hakan Ozcelik. "Toxic Decision Processes: A Study of Emotion and Organizational Decision-making." *Organization Science* 15, no. 4 (2004). 384–391.

82. Pellegrino, Edmund D., H. Tristram Engelhardt, and Fabrice Jotterand. *The philosophy of medicine reborn: a Pellegrino reader*. Notre Dame: University of Notre Dame Press, 2011. 255–264.

83. Devettere, Raymond J. *Practical decision-making in health care ethics: cases, concepts, and the virtue of prudence*. Washington, D.C.: (Georgetown University Press, 2016), 14–17.

84. Runciman, Bill, Alan Merry, and Merrilyn Walton. *Safety and ethics in healthcare a guide to getting it right*. Aldershot, England: Ashgate, 2007. 135–154.

85. Dunn, Barnaby D., Hannah C. Galton, Ruth Morgan, Davy Evans, Clare Oliver, Marcel Meyer, Rhodri Cusack, Andrew D. Lawrence, and Tim Dalgleish. "Listening to Your Heart." *Psychological Science* 21, no. 12 (2010).1842.

86. Chidwick, Paula, Karen Faith, Dianne Godkin, and Laurie Hardingham. "Clinical education of ethicists: the role of a clinical ethics fellowship." *BMC Medical Ethics* 5, no. 1 (2004). 1–8.

87. Emanuel, Ezekiel J., Christine C. Grady, Robert A. Crouch, Reidar K. Lie, Franklin G. Miller, and David D. Wendler, eds, "The Nuremburg Code," In *The Oxford textbook of clinical research ethics*, (Oxford University Press, 2008), 136–138.

88. Emanuel, Ezekiel J., Christine C. Grady, Robert A. Crouch, Reidar K. Lie, Franklin G. Miller, and David D. Wendler, eds, "The Nuremburg Code," In *The Oxford textbook of clinical research ethics*, (Oxford University Press, 2008), 137.

89. Emanuel, Ezekiel J., Christine C. Grady, Robert A. Crouch, Reidar K. Lie, Franklin G. Miller, and David D. Wendler, eds, "The Nuremburg Code," In *The Oxford textbook of clinical research ethics*, (Oxford University Press, 2008), 138–139.

90. Emanuel, Ezekiel J., Christine C. Grady, Robert A. Crouch, Reidar K. Lie, Franklin G. Miller, and David D. Wendler, eds, "The Nuremburg Code," In *The Oxford textbook of clinical research ethics*, (Oxford University Press, 2008), 138–139.

91. Emanuel, Ezekiel J., Christine C. Grady, Robert A. Crouch, Reidar K. Lie, Franklin G. Miller, and David D. Wendler, eds, "The Nuremburg Code," In *The Oxford textbook of clinical research ethics*, (Oxford University Press, 2008), 139.

92. Emanuel, Ezekiel J., Christine C. Grady, Robert A. Crouch, Reidar K. Lie, Franklin G. Miller, and David D. Wendler, eds, "The Nuremburg Code," In *The Oxford textbook of clinical research ethics*, (Oxford University Press, 2008), 138.

93. Emanuel, Ezekiel J., Christine C. Grady, Robert A. Crouch, Reidar K. Lie, Franklin G. Miller, and David D. Wendler, eds, "The Nuremburg Code," In *The Oxford textbook of clinical research ethics*, (Oxford University Press, 2008), 139.

94. Emanuel, Ezekiel J., Christine C. Grady, Robert A. Crouch, Reidar K. Lie, Franklin G. Miller, and David D. Wendler, eds, "The Nuremburg Code," In *The Oxford textbook of clinical research ethics*, (Oxford University Press, 2008), 139.

95. Emanuel, Ezekiel J., Christine C. Grady, Robert A. Crouch, Reidar K. Lie, Franklin G. Miller, and David D. Wendler, eds, "The Nuremburg Code," In *The Oxford textbook of clinical research ethics*, (Oxford University Press, 2008), 139.

96. Emanuel, Ezekiel J., Christine C. Grady, Robert A. Crouch, Reidar K. Lie, Franklin G. Miller, and David D. Wendler, eds, "The Nuremburg Code," In *The Oxford textbook of clinical research ethics*, (Oxford University Press, 2008), 139.

97. Emanuel, Ezekiel J., Christine C. Grady, Robert A. Crouch, Reidar K. Lie, Franklin G. Miller, and David D. Wendler, eds, "The Nuremburg Code," In *The Oxford textbook of clinical research ethics*, (Oxford University Press, 2008), 140.

98. Emanuel, Ezekiel J., Christine C. Grady, Robert A. Crouch, Reidar K. Lie, Franklin G. Miller, and David D. Wendler, eds, "The Nuremburg Code," In *The Oxford textbook of clinical research ethics*, (Oxford University Press, 2008), 140.

99. Emanuel, Ezekiel J., Christine C. Grady, Robert A. Crouch, Reidar K. Lie, Franklin G. Miller, and David D. Wendler, eds, "The Nuremburg Code," In *The Oxford textbook of clinical research ethics*, (Oxford University Press, 2008), 140.

100. Beauchamp, Tom L., and James F. Childress. *Principles of Biomedical Ethics*. 5th ed. (New York, N.Y: Oxford University Press, 2013), 120–121.

101. Beauchamp, Tom L., and James F. Childress. *Principles of Biomedical Ethics*. 5th ed. (New York, N.Y: Oxford University Press, 2013), 101.

102. Beauchamp, Tom L., and James F. Childress. *Principles of Biomedical Ethics*. 5th ed. (New York, N.Y: Oxford University Press, 2013), 101–102.

103. Beauchamp, Tom L., and James F. Childress. *Principles of Biomedical Ethics*. 5th ed. (New York, N.Y: Oxford University Press, 2013), 101–102.

104. Beauchamp, Tom L., and James F. Childress. *Principles of Biomedical Ethics*. 5th ed. (New York, N.Y: Oxford University Press, 2013), 153.

105. Beauchamp, Tom L., and James F. Childress. *Principles of Biomedical Ethics*. 5th ed. (New York, N.Y: Oxford University Press, 2013), 154.

106. Beauchamp, Tom L., and James F. Childress. *Principles of Biomedical Ethics*. 5th ed. (New York, N.Y: Oxford University Press, 2013), 154–156.

107. Youngberg, Barbara J. *Patient safety handbook*. (Jones & Bartlett Publishers, 2012), 2–22.

108. Beauchamp, Tom L., and James F. Childress. *Principles of Biomedical Ethics*. 5th ed. (New York, N.Y: Oxford University Press, 2013), 150–151.

109. Beauchamp, Tom L., and James F. Childress. *Principles of Biomedical Ethics*. 5th ed. (New York, N.Y: Oxford University Press, 2013), 150–151.

110. Beauchamp, Tom L., and James F. Childress. *Principles of Biomedical Ethics*. 5th ed. (New York, N.Y: Oxford University Press, 2013), 151.

111. Beauchamp, Tom L., and James F. Childress. *Principles of Biomedical Ethics*. 5th ed. (New York, N.Y: Oxford University Press, 2013), 361.

112. Beauchamp, Tom L., and James F. Childress. *Principles of Biomedical Ethics*. 5th ed. (New York, N.Y: Oxford University Press, 2013), 361.

113. Beauchamp, Tom L., and James F. Childress. *Principles of Biomedical Ethics*. 5th ed. (New York, N.Y: Oxford University Press, 2013), 361.

114. Beauchamp, Tom L., and James F. Childress. *Principles of Biomedical Ethics*. 5th ed. (New York, N.Y: Oxford University Press, 2013), 363.

115. Beauchamp, Tom L., and James F. Childress. *Principles of Biomedical Ethics*. 5th ed. (New York, N.Y: Oxford University Press, 2013), 363.

116. Beauchamp, Tom L., and James F. Childress. *Principles of Biomedical Ethics*. 5th ed. (New York, N.Y: Oxford University Press, 2013), 363.

117. Beauchamp, Tom L., and James F. Childress. *Principles of Biomedical Ethics*. 5th ed. (New York, N.Y: Oxford University Press, 2013), 363.

118. Beauchamp, Tom L., and James F. Childress. *Principles of Biomedical Ethics*. 5th ed. (New York, N.Y: Oxford University Press, 2013), 364.

119. Beauchamp, Tom L., and James F. Childress. *Principles of Biomedical Ethics*. 5th ed. (New York, N.Y: Oxford University Press, 2013), 366.

120. Beauchamp, Tom L., and James F. Childress. *Principles of Biomedical Ethics*. 5th ed. (New York, N.Y: Oxford University Press, 2013), 361.

121. Beauchamp, Tom L., and James F. Childress. *Principles of Biomedical Ethics*. 5th ed. (New York, N.Y: Oxford University Press, 2013), 366–367.

122. Beauchamp, Tom L., and James F. Childress, "Moral Theories," In *Principles of Biomedical Ethics*. 5th ed. (New York, N.Y.: Oxford University Press, 2001), 366–367.

123. Beauchamp, Tom L., and James F. Childress, "Moral Theories," In *Principles of Biomedical Ethics*. 5th ed. (New York, N.Y.: Oxford University Press, 2001), 363–364.

124. Beauchamp, Tom L., and James F. Childress, "Moral Theories," In *Principles of Biomedical Ethics*. 5th ed. (New York, N.Y.: Oxford University Press, 2001), 354.

125. Beauchamp, Tom L., and James F. Childress, "Moral Theories," In *Principles of Biomedical Ethics*. 5th ed. (New York, N.Y.: Oxford University Press, 2001), 354.

126. Beauchamp, Tom L., and James F. Childress, "Moral Theories," In *Principles of Biomedical Ethics*. 5th ed. (New York, N.Y.: Oxford University Press, 2001), 354.

127. Beauchamp, Tom L., and James F. Childress, "Moral Theories," In *Principles of Biomedical Ethics*. 5th ed. (New York, N.Y.: Oxford University Press, 2001), 355.

128. Salloch, Sabine, Jan Schildmann, and Jochen Vollmann. "Empirical Research in Medical Ethics: How Conceptual Accounts on Normative-empirical Collaboration May Improve Research Practice." *BMC Med Ethics* 13, no. 1, (2012): 23–32.

129. Beauchamp, Tom L., and James F. Childress, "Moral Theories," In *Principles of Biomedical Ethics*. 5th ed. (New York, N.Y.: Oxford University Press, 2001), 363.

130. Beauchamp, Tom L., and James F. Childress, "Moral Theories," In *Principles of Biomedical Ethics*. 5th ed. (New York, N.Y.: Oxford University Press, 2001), 357.

131. Beauchamp, Tom L., and James F. Childress, "Moral Theories," In *Principles of Biomedical Ethics*. 5th ed. (New York, N.Y.: Oxford University Press, 2001), 357.

132. Beauchamp, Tom L., and James F. Childress, "Moral Theories," In *Principles of Biomedical Ethics*. 5th ed. (New York, N.Y.: Oxford University Press, 2001), 359.

133. Chadwick, Ruth F., Henk Ten Have, and Eric M. Meslin. "Research Ethics," In *The SAGE Handbook of Health Care Ethics*. (Los Angeles, California: SAGE, 2011), 313.

134. Chadwick, Ruth F., Henk Ten Have, and Eric M. Meslin. "Research Ethics," In *The SAGE Handbook of Health Care Ethics*. (Los Angeles, California: SAGE, 2011), 313.

135. Chadwick, Ruth F., Henk Ten Have, and Eric M. Meslin. "Research Ethics," In *The SAGE Handbook of Health Care Ethics*. (Los Angeles, California: SAGE, 2011), 313.

136. Chadwick, Ruth F., Henk Ten Have, and Eric M. Meslin. "Research Ethics," In *The SAGE Handbook of Health Care Ethics*. (Los Angeles, California: SAGE, 2011), 313.

137. Chadwick, Ruth F., Henk Ten Have, and Eric M. Meslin. "Research Ethics," In *The SAGE Handbook of Health Care Ethics*. (Los Angeles, California: SAGE, 2011), 313.

138. Chadwick, Ruth F., Henk Ten Have, and Eric M. Meslin. "Research Ethics," In *The SAGE Handbook of Health Care Ethics*. (Los Angeles, California: SAGE, 2011), 313.

139. Chadwick, Ruth F., Henk Ten Have, and Eric M. Meslin. "Research Ethics," In *The SAGE Handbook of Health Care Ethics*. (Los Angeles, California: SAGE, 2011), 313.

140. Chadwick, Ruth F., Henk Ten Have, and Eric M. Meslin. "Research Ethics," In *The SAGE Handbook of Health Care Ethics*. (Los Angeles, California: SAGE, 2011), 313.

141. Chadwick, Ruth F., Henk Ten Have, and Eric M. Meslin. "Advance Directives," In *The SAGE Handbook of Health Care Ethics*. (Los Angeles, California: SAGE, 2011), 219–223.

142. Chadwick, Ruth F., Henk Ten Have, and Eric M. Meslin. "Advance Directives," In *The SAGE Handbook of Health Care Ethics*. (Los Angeles, California: SAGE, 2011), 220–224.

143. Chadwick, Ruth F., Henk Ten Have, and Eric M. Meslin. "Advance Directives," In *The SAGE Handbook of Health Care Ethics*. (Los Angeles, California: SAGE, 2011), 224.

144. Chadwick, Ruth F., Henk Ten Have, and Eric M. Meslin. "Advance Directives," In *The SAGE Handbook of Health Care Ethics*. (Los Angeles, California: SAGE, 2011), 224.
145. Chadwick, Ruth F., Henk Ten Have, and Eric M. Meslin. "Advance Directives," In *The SAGE Handbook of Health Care Ethics*. (Los Angeles, California: SAGE, 2011), 224.
146. Beauchamp, Tom L., and James F. Childress. "Moral Status," In *Principles of Biomedical Ethics*. 5th ed. (New York, N.Y.: Oxford University Press, 2001), 90–91.
147. Beauchamp, Tom L., and James F. Childress. "Moral Status," In *Principles of Biomedical Ethics*. 5th ed. (New York, N.Y.: Oxford University Press, 2001), 90–91.
148. Beauchamp, Tom L., and James F. Childress. "Moral Status," In *Principles of Biomedical Ethics*. 5th ed. (New York, N.Y.: Oxford University Press, 2001), 90–91.
149. Beauchamp, Tom L., and James F. Childress. "Moral Status," In *Principles of Biomedical Ethics*. 5th ed. (New York, N.Y.: Oxford University Press, 2001), 90–91.
150. Chadwick, Ruth F., Henk Ten Have, and Eric M. Meslin. "Advance Directives," In *The SAGE Handbook of Health Care Ethics*. (Los Angeles, California: SAGE, 2011), 244–245.
151. Beauchamp, Tom L., and James F. Childress. "Moral Status," In *Principles of Biomedical Ethics*. 5th ed. (New York, N.Y.: Oxford University Press, 2001), 30–32.
152. Beauchamp, Tom L., and James F. Childress. "Moral Status," In *Principles of Biomedical Ethics*. 5th ed. (New York, N.Y.: Oxford University Press, 2001), 30–32.
153. Beauchamp, Tom L., and James F. Childress. "Moral Status," In *Principles of Biomedical Ethics*. 5th ed. (New York, N.Y.: Oxford University Press, 2001), 30–32.
154. Beauchamp, Tom L., and James F. Childress. "Moral Status," In *Principles of Biomedical Ethics*. 5th ed. (New York, N.Y.: Oxford University Press, 2001), 32–35.
155. Beauchamp, Tom L., and James F. Childress. "Moral Status," In *Principles of Biomedical Ethics*. 5th ed. (New York, N.Y.: Oxford University Press, 2001), 30–35.
156. Beauchamp, Tom L., and James F. Childress. "Moral Status," In *Principles of Biomedical Ethics*. 5th ed. (New York, N.Y.: Oxford University Press, 2001), 35.
157. Beauchamp, Tom L., and James F. Childress. "Moral Status," In *Principles of Biomedical Ethics*. 5th ed. (New York, N.Y.: Oxford University Press, 2001), 37–39.
158. Beauchamp, Tom L., and James F. Childress. "Moral Status," In *Principles of Biomedical Ethics*. 5th ed. (New York, N.Y.: Oxford University Press, 2001), 39.
159. Beauchamp, Tom L., and James F. Childress. "Moral Status," In *Principles of Biomedical Ethics*. 5th ed. (New York, N.Y.: Oxford University Press, 2001), 39–40.
160. Beauchamp, Tom L., and James F. Childress. "Moral Status," In *Principles of Biomedical Ethics*. 5th ed. (New York, N.Y.: Oxford University Press, 2001), 42–44.
161. Beauchamp, Tom L., and James F. Childress. "Moral Status," In *Principles of Biomedical Ethics*. 5th ed. (New York, N.Y.: Oxford University Press, 2001), 44.
162. Pellegrino, Edmund D., and David C. Thomasma. "A Philosophical Reconstruction of Medical Morality." In *A Philosophical Basis of Medical Practice: Toward a Philosophy and Ethic of the Healing Professions*. New York: Oxford University Press, 1981. 192.
163. Pellegrino, Edmund D., and David C. Thomasma. "A Philosophical Reconstruction of Medical Morality." In *A Philosophical Basis of Medical Practice: Toward a Philosophy and Ethic of the Healing Professions*. New York: Oxford University Press, 1981. 192.
164. Pellegrino, Edmund D., and David C. Thomasma. "A Philosophical Reconstruction of Medical Morality." In *A Philosophical Basis of Medical Practice: Toward a Philosophy and Ethic of the Healing Professions*. New York: Oxford University Press, 1981. 192.
165. Pellegrino, Edmund D., and David C. Thomasma. "A Philosophical Reconstruction of Medical Morality." In *A Philosophical Basis of Medical Practice: Toward a Philosophy and Ethic of the Healing Professions*. New York: Oxford University Press, 1981. 192.
166. Pellegrino, Edmund D., and David C. Thomasma. "A Philosophical Reconstruction of Medical Morality." In *A Philosophical Basis of Medical Practice: Toward a Philosophy and Ethic of the Healing Professions*. New York: Oxford University Press, 1981. 192. 193.

167. Pellegrino, Edmund D., and David C. Thomasma. "A Philosophical Reconstruction of Medical Morality." In *A Philosophical Basis ofMedical Practice: Toward a Philosophy and Ethic of the Healing Professions*. New York: Oxford University Press, 1981. 192–193.
168. Pellegrino, Edmund D., and David C. Thomasma. "A Philosophical Reconstruction of Medical Morality." In *A Philosophical Basis of Medical Practice: Toward a Philosophy and Ethic of the Healing Professions*. New York: Oxford University Press, 1981. 194.
169. Pellegrino, Edmund D., and David C. Thomasma. "A Philosophical Method." In *A Philosophical Basis of Medical Practice: Toward a Philosophy and Ethic of the Healing Professions*. New York: Oxford University Press, 1981. 39.
170. Sugarman, Jeremy "Philosophy: Ethical Principles and Common Morality," In Methods in Medical Ethics, (Washington, D.C.: Georgetown University Press, 2010), 40–41.
171. Sugarman, Jeremy "Philosophy: Ethical Principles and Common Morality," In Methods in Medical Ethics, (Washington, D.C.: Georgetown University Press, 2010), 40–41.
172. Sugarman, Jeremy "Philosophy: Ethical Principles and Common Morality," In Methods in Medical Ethics, (Washington, D.C.: Georgetown University Press, 2010), 40–41.
173. Sugarman, Jeremy "Philosophy: Ethical Principles and Common Morality," In Methods in Medical Ethics, (Washington, D.C.: Georgetown University Press, 2010), 42–43.
174. Sugarman, Jeremy "Philosophy: Ethical Principles and Common Morality," In Methods in Medical Ethics, (Washington, D.C.: Georgetown University Press, 2010), 65.
175. Sugarman, Jeremy "Philosophy: Ethical Principles and Common Morality," In Methods in Medical Ethics, (Washington, D.C.: Georgetown University Press, 2010), 65.
176. Sugarman, Jeremy "Philosophy: Ethical Principles and Common Morality," In Methods in Medical Ethics, (Washington, D.C.: Georgetown University Press, 2010), 65.
177. Sugarman, Jeremy "Philosophy: Ethical Principles and Common Morality," In Methods in Medical Ethics, (Washington, D.C.: Georgetown University Press, 2010), 65.
178. Sugarman, Jeremy "Philosophy: Ethical Principles and Common Morality," In Methods in Medical Ethics, (Washington, D.C.: Georgetown University Press, 2010), 65.
179. MacIntyre, Alasdair C. "Aristotle's Account of the Virtues." In *After Virtue: A Study in Moral Theory*. 3rd ed. Notre Dame, Ind.: University of Notre Dame Press, 2008. 146–147.
180. MacIntyre, Alasdair C. "Aristotle's Account of the Virtues." In *After Virtue: A Study in Moral Theory*. 3rd ed. Notre Dame, Ind.: University of Notre Dame Press, 2008.
181. MacIntyre, Alasdair C. "Aristotle's Account of the Virtues." In *After Virtue: A Study in Moral Theory*. 3rd ed. Notre Dame, Ind.: University of Notre Dame Press, 2008. 148–149.
182. MacIntyre, Alasdair C. "Aristotle's Account of the Virtues." In *After Virtue: A Study in Moral Theory*. 3rd ed. Notre Dame, Ind.: University of Notre Dame Press, 2008. 150–157.
183. MacIntyre, Alasdair C. "Aristotle's Account of the Virtues." In *After Virtue: A Study in Moral Theory*. 3rd ed. Notre Dame, Ind.: University of Notre Dame Press, 2008. 150–157.
184. MacIntyre, Alasdair C. "Aristotle's Account of the Virtues." In *After Virtue: A Study in Moral Theory*. 3rd ed. Notre Dame, Ind.: University of Notre Dame Press, 2008. 160–162.
185. MacIntyre, Alasdair C. "Aristotle's Account of the Virtues." In *After Virtue: A Study in Moral Theory*. 3rd ed. Notre Dame, Ind.: University of Notre Dame Press, 2008. 160–162.
186. MacIntyre, Alasdair C. "From the Virtues to Virtue and after Virtue." In *After Virtue: A Study in Moral Theory*. 3rd ed. Notre Dame, Ind.: University of Notre Dame Press, 2008. 226–227.
187. MacIntyre, Alasdair C. "From the Virtues to Virtue and after Virtue." In *After Virtue: A Study in Moral Theory*. 3rd ed. Notre Dame, Ind.: University of Notre Dame Press, 2008. 226–227–231.

# Chapter 4
# Organizational Ethics and Residency Requirements for Clinical Ethics Consultation

The existence of ethics in non-healthcare organizations is more debated among professional circles than in medicine. The current difficulties in determining the role of ethics, including its limitations, stem from a misuse of ethics education and a misunderstanding of vocational ethics in general [1]. By identifying ethics in professional practice, both the role and scope of ethics at the organizational level become clearer. The concepts of professional character, integrity, and stewardship play vital roles in the betterment of an organization [2]. These characteristic elements within organizational ethics are both indicative and obligatory of a moral institution [3].

Two commonly accepted premises concerning human rights include: (1) All human life is sacred and (2) All human beings are ends within themselves. While the former premise receives support and solidification from various religious doctrines and contemporary standards of care, the latter derives from the Kantian conception of human rights [4]. Since every human being is sacred, individuals or an organization of individuals should accept or reject decisions based on the effect on other human beings. While decision-making on the individual level accompanies an array of considerations, organizational decision-making is a far more onerous task due to the varying perspectives that arise from a collective of individuals. For the sake of this discussion and the role of ethics in health care, hospitals serve as the primary organizational institution where ethical decision-making occurs [5].

Organizational ethics pertains to a category of notions that wed institutional goals with corporate character. These notions derive from tactical business plans that amalgamate compliance, legality, and regulations. However, the scope of organizational integrity and stewardship must extend beyond these bare facets of institutional functionality. Health care organizations must seek a greater understanding of institutional leadership by expanding their regulations and corporate requirements beyond the facets mentioned above of organizational structures [6]. While the expansion of a health care organization's ethical scope must encompass the ASBH's standards for policy positions and behaviors regarding patients, employees, and partners, the primary goal of a health care organization must lie in upholding the

J. T. Bertino, *Clinical Ethics for Consultation Practice*, https://doi.org/10.1007/978-3-030-90182-0_4

quintessence of virtue, ethical principles, value-based decision-making, and upholding an appropriate code of conduct. In this respect, a fluid relationship between these goals inevitably yields a virtuous organization [7].

Understanding professional character and professional ethics become apparent if a practitioner's moral responsibilities demonstrate continuity with their affiliated organization's responsibilities. With the development of corporate models for health care institutions, physicians have adopted a stakeholder role that permeates the barrier of ethical practice [8]. Physicians have steadily become involved in healthcare institutions' financial interests by investing in ambulatory surgery centers, radiology clinics, and prenatal hospitals [9]. Furthermore, the same physicians financially profit from the services provided by an institution. While physicians require compensation for their services, the financial benefits rendered from services have created an ominous and pervasive relationship between healthcare providers and patient populations. Since the same services provided by physicians are not accessible without an institutional venue, individuals are left with no recourse but to engage with an organization that weds its practices and financial motivations with one another. Since the financial motivations behind health services inherently bias health care providers, human beings are at risk of becoming means to a financial end rather than ends in themselves [10].

The current state of financial bias and financial relationships within health care organizations create a significant challenge for health care ethics. Although a humanitarian or altruistic aim is often not always the organization's goal, maintaining human dignity and respect throughout an organization's practice is paramount to the goods and services provided. However, health care institutions inherently the dignity of human beings in their practice. Assessing the dichotomy between services provided and respect is tenable if the moral agency of hospitals and health care teams receives attention. This task is difficult at the institutional level because fiscal considerations and self-interest have saturated the moral lens of individuals [11]. While financial concerns are unavoidable, the overall intentions of an organization ought to hold ethical weight, especially when dealing with individuals who require a service. By examining the model of a virtuous practitioner, appropriate methods of organizational practice become tenable. Furthermore, adopting a philosophical basis for medical practice aids in mending the shortcomings of organizational practices [12].

The scope of an organization depends on the intentions and goals of the institution. The intricacies of a health care organization's services dictate an organization's intentions. However, health care institutions like hospitals, nursing home facilities and other relevant health care organizations work directly with human beings in need of aid that they cannot receive independently. Due to the nature of their work, health care organizations accompany an inherent obligation to their clientele [13]. While other institutions may strictly deal with individuals' livelihood and financial interests, health care organizations assess and manage inalienable aspects of human life—namely, the human body and the ailments afflict biological structures. Since the services that health care organizations provide deal specifically with the flourishing of human life, they inseparably wed themselves to standards that must uphold professional integrity and stewardship across their practice [14].

A virtuous organization demonstrates its tact to the public through its practitioners. Simply put, an organization is virtuous if its employees and involved stakeholders reciprocally perform their functions. At the heart of organizational integrity lies a premise of completion. Specifically, the concept of integrity attempts to bridge the gap between an organization's public and private character [15]. Synonymous with the late Hannah Arendt's work in *The Human Condition*, the exposure of an individual's public and private life is indicative of one's inherent character [16]. However, when dealing with an organization, the private affairs of any given institution must be reflected outward to the public. This process requires organizational integrity. In many respects, organizational integrity promises the public who directly associate with a health care intuition. While divulging every detail of a health care organization to the public is not possible, it is the responsibility of the organization to assure and maintain its goals and obligations to the people the organization serves. In this respect, a tremendous amount of trust embeds itself within the concept of organizational integrity [17]. The organization must trust the public insofar as the public recognizes the limitations and scope of a health care organization, while the public trusts the health care organization to perform its duties in full service and concerning its patients [18].

While integrity pertains to the relationship an organization upholds between its practices and the public, stewardship involves the obligatory maintenance required to uphold a trusting and influential institution. Since health care institutions are in the business of health and well-being of others, stewardship becomes the conceptualized manifestation of moral partnership and duty that inherently ties to the mission of a health care institution [19]. Furthermore, stewardship inspires virtue within an organization insofar as the work conducted by an organization seeks to better the community it serves. Stewardship bolsters a health care organization's commitment to a community by clearly delineating goals that involve care, ethical practice, fairness, and justice [20].

The mission and values of a health care institution revolve around the betterment of human health and human flourishing. While the core values and mission of a health care institution vary, it is essential to balance the goals of a health care institution with the impending necessities of a free-market economy [21]. In this respect, the integrity and stewardship of an institution challenge the pressures of financial gain with a mission of healing and service. Promoting a health care institution's mission and core values requires a thorough understanding and establishment of a mission statement that emphasizes core values [22].

A health care institution's mission statement should exercise brevity but also power in its language. For instance, the mission statement of Trinity Health, a Michigan-based Catholic institution, reads, "[w]e, Trinity Health, serve together in the spirit of the Gospel as a compassionate and transforming healing presence within our communities." [23] This mission statement is effective for several reasons. First, it indicates the institution's role and functionality. Second, the statement clearly illustrates the organization's Catholic identity. Finally, the statement attempts to demonstrate that the mission goes beyond the duties of a health care system by penetrating various facets of human well-being [24].

The efficacy and prowess of a workforce are demonstrative of a well-formulated mission statement. Employees of a healthcare organization should refer to their institution's mission statement to remind themselves and others that their organization's role extends far beyond a single employee's duties. With every subsequent project, report, procedure, consultation, or ask a health care organization provides, institutional understanding of the boundlessness that a mission statement entails can significantly impact the workplace's conduct and atmosphere [25]. While core values of a given health care institution curtail the specific duties of that institution, the underlying function, and values that an organization formulates base themselves on stewardship and integrity. Examples of core values include reverence, commitment to those who are poor, and justice. Core values should become part of an organization's corpus if they demonstrate relevance and serve as catalysts for expediting the institution's mission. For instance, a catholic health care organization may possess a core value of reverence due to its basis in honoring the sacredness and dignity of every person. Reverence is a core value that should be on every health care professional's mind when seeing a new patient and thus fits well into the mission and values of a healing enterprise [26].

Core values like justice and stewardship demonstrate their rootedness in an organization's mission due to their efficacy in helping patient populations and corporate decision-making. However, this analysis indicates the relevance of virtue throughout an organization. In this respect, a health care organization's core values and mission should align with, or become identical, with moral virtue [27]. In doing so, identification of virtue becomes an imminent aspect of organizational work. The unavoidable and inherent nature of identifying virtues that double as an organization's core values prompts employees to develop and hone analytic moral reasoning skills throughout their practice. While the virtue of justice focuses on fostering proper relationships to promote the common good, stewardship honors the heritage of these proper relationships. Subsequently, it holds those who foster the same relationships accountable for the human, financial, and natural resources entrusted within a health care organization [28].

Although integrating core values and mission integration within an organization weighs heavily in moral theory, the processes, and tasks involved within integrating stewardship integrity further associate with ethics and corporate compliance. Insofar as virtue ethics serves as its foundation, mission integration must oversee all organizational processes, including strategic planning and implementing goals, conducting contract agreements, and overseeing partnership relationships with other organizations [29].

The core values discussed above demonstrate the administrative scaffold that all health care organizations should acknowledge. Without these grounding principles and values, the educational possibilities involved in identifying virtue and promoting analytical moral reasoning skills thwarts the quality and relevant discourse necessary for stewardship and integrity. Stewardship and integrity are at the foundation of organizational health care. These foundational aspects of health care aid the contemporary development of virtuous care and serve as investment strategies for future organizations and practitioners. Continuing the mission of a virtuous organization is a tenable feat in a competitive market if the goals of an organization focus on the betterment of individuals and the promotion of human well-being [30]. While a lasting and effective financial infrastructure is necessary for an organization's function, it is still a secondary component when juxtaposed with the importance of an organization's moral fortitude. Structuring an entire health care organization around a mission of stewardship and ethical integrity solidifies a promising future and aids in establishing a virtuous institution [31].

Although integrity pertains to a relationship grounded in trust, various characteristics contribute to the overall composition of organizational integrity. Furthermore, many of these contributions are moral virtues that directly pertain to the betterment of an institution's functioning and its contributions to human flourishing. These characteristics include, among others, courage, honesty, responsibility, accountability, justice, respect, humility, and commitment [32]. The overlap created between these characteristics of moral virtue contains a two-fold benefit for organizational integrity and stewardship. First, acknowledging and practicing these virtues bolster the moral fortitude of the individuals responsible for a well-functioning organization. Second, these virtues aid in solidifying a relationship of moral integrity between a healthcare organization and the populations the organization aids. By identifying virtue in professional practice, health care organizations are subsequently able to develop methods of analytical moral decision-making throughout their practice [33].

Bolstering moral fortitude in organizations is a widely recognized need amongst ethicists and moral scholars alike [34]. The enhancement of an organization's understanding of moral virtue and professional practice yields significant corporate progress. Additionally, a morally sound organization enhances employee's understanding of an institution's mission and aids in solidifying a trusting relationship between an organization and its clientele. This analysis demonstrates this relationship as organizational integrity later, yet the moral elements that contribute to this relationship manifest as moral virtues. Synonymous with Aristotelian virtue ethics discussed previously, one may identify moral virtues in organizations by acknowledging virtue as a mean between two possible extremes. While an array of virtues manifest in a health care organization, courage and justice appear frequently. For the sake of this discussion, these two virtues encompass relevant facets of analytical moral reasoning. However, it is essential to acknowledge the utility of virtuous application across a given health care organization [35].

The seemingly broad nature of virtues may initially dissuade facilitation efforts in organizational ethics, yet the method of identifying virtue is the focal intention of this analysis. Identifying instances of courage and justice, for instance, within an organization promotes lasting skills and aid in the development and understanding of virtue in practice [36].In health care organizations, courage and justice are embodiments of good decision-making, facilitation, conflict resolution, and character development. The analytic moral reasoning skills developed from identifying moral virtue in health care organizations forge integrity and stewardship to its patients and develops an analytical method for health care professionals and health care organizations [37].

Excesses and deficiencies of moral virtues manifest differently. Concerning moral courage, 'cowardice' represents virtuous deficiency while 'rashness' represents virtuous excess. Regarding Justice, excesses and deficiencies result in imbalances of power. Deficiencies and excesses of virtue frequently appear in health care ethics consultations and are frequently the crux of ethical discrepancies. However, organizational ethics cases possess a unique character within their hierarchical status [38]. The shifting of roles, responsibilities, and statuses within an organizational structure can easily provoke power imbalances. However, the ideal organizational structure remains fluid to accommodate policy adjustments [39]. An organizational hierarchy functions virtuously if the changes conduct with the interest of the individuals affected by the organization [40].

When an organizational ethics issue arises in health care, the result can often trace to an excess or deficiency in virtue. These excesses and deficiencies in virtue, although problematic, present opportunities for moral agents to recognize the differences between shortcomings of moral fortitude and opportunities for ethical deliberation. To illustrate the difference between excess and deficiency in virtue, consider the following situation: Two doctors perform a routine outpatient procedure on a 34-year-old female. Although familiar with the procedure, both doctors use new equipment with no training or possess organizational authorization. Additionally, a salesperson from the equipment manufacturer assists the procedure by operating the equipment while the doctors perform the procedure. A nurse expresses concern regarding the apparent violations of organizational policy, but the physicians disregard the nurse's concerns. During the procedure, the patient dies [41].

The above situation contains an array of ethical discrepancies, including violations of organizational policy. However, the most threatening ethical concern regarding this case involves an overstepping of expertise. The physicians responsible for the patient's care did not act courageously because they overstepped their expertise and did not exercise poise. Although both physicians may have felt confident in their abilities, a critical failure occurred during the procedure due to a lack of analytical moral reasoning [42]. Had both physicians taken a moment to identify what actions align with virtuous medical practice, they may have reconsidered making a rash decision, namely, performing a procedure with minimal equipment experience with an unauthorized individual. Alternatively, a deficiency of moral courage in this instance may have resulted in the act of cowardice. Cowardice can manifest in various ways concerning this clinical case, i.e., covering the mistake up, blaming the malpractice on another individual, or lying on electronic medical records to protect one's reputation [43].

While the physicians in the case mentioned above exhibit vast misrepresentations of virtue, the healthcare organization responsible for the conduct of the physicians must identify their inappropriate actions and develop institutional policies to aid in thwarting these detrimental instances in the future. Furthermore, the health care organization can bolster underappreciate and valuable healthcare professionals' voices that do not receive proper attention due to hierarchical structures, i.e., nurse-physician relationships. While an institution's policies must uphold the same standards of stewardship and integrity discussed earlier in this analysis, organizations must also abide by a standard of practices that reinforces these standards [44]. While this aspect of organizational ethics seemingly belongs in the realm of compliance, the two disciplines are complementary insofar as they intend to reinforce organizational policy for the sake of the organization's mission. Although addressing the importance of wrongness, sanctions, and conflict resolution receive attention in this analysis, understanding how to identify virtue in medical practice aids health care organizations in developing a method that facilitates analytic moral reasoning skills. These skills saturate the various components of a health care organization and aid in preventing the metastasis of excessive and deficient instances of virtue or organizational wrong-doing [45].

Recognizing how virtue manifests in health care organizations is far more tenable if moral agents can recognize deficiencies and excesses of virtue. Moral agents develop analytical moral reasoning skills that shape an organization's moral fortitude, character, and stakeholders. Shaping the moral fortitude and character of a health care organization promotes the foundational dimensions of ethical practices like stewardship, integrity, and patient flourishing [46]. However, it is essential to note that the betterment of an organization results from the effective implementation and practice of analytic moral reasoning skills. If implemented efficiently and effectively, the subsequent cultivation of moral fortitude and character follows suit within a health care organization. The proper implementation of analytic moral reasoning skills shapes moral fortitude and character because exercising analytic moral reasoning skills involves active participation in ethical practices [47].

Leslie Sekerka, Richard Bagozzi, and Richard Charnigo demonstrate the effectiveness of virtuous practice by coining the phrase "Professional Moral Courage" (PMC). According to the authors, PMC is both measurable and teachable. In this respect, the difficulties associated with promoting virtue in health care organizations are moot due to the quantifiable and teachable nature of virtue [48]. However, measuring and teaching PMC also begs the question: to what degree does measuring and teaching PMC to promote moral decision-making, positive organizational scholarship, and organizational ethics? Examining PMC as a competency or a central theme that can apply to ethical situations and directly influence action is necessary. A competency is a general description or overview of skills used to perform tasks in a workplace effectively. Synonymous with the ASBH's core competencies, the same stringent criteria should apply to virtue in health care organizations and clinical care alike [49]. However, the difference between practical core competencies and shaping virtues as competencies lie in their skillsets. While the ASBH's core competencies intend to guide and aid facilitation efforts of clinical ethics consultants and other health care professionals, virtuous competencies shape and develop analytic moral reasoning skills for clinicians, ethics consultants, and all health care administrators [50].

Sekerka, Bagozzi and Charnigo's confidence in the measurability and teachable components of PMC is promising for various reasons. First, deeming virtue as a competency opens various avenues for organizational leadership and ethical promotion in health care. Recognizing the necessity of understanding virtue as competencies across a health care organization promotes confidence in moral agents. Rather than questioning whether a moral agent should deliberate between two extremes, moral agents subsequently accept the process of seeking moral means as obligations rather than mere options [51]. Second, accepting virtues as practical competencies shapes moral attitudes and character in a health care organization. While practical competencies demonstrate viable protocols in ethical practice, competencies in virtue encourage moral agents to actualize and pursue moral goals. These goals entail prudence, honesty, and even other virtuous facets like justice [52]. Finally, upholding virtuous competencies aid in the process of identifying virtue in health care settings.

This analysis has stressed the importance of identifying virtue to form and honing analytic moral reasoning skills. These skills are subsequent byproducts that result from actively participating in virtue-identification. However, analytic moral reasoning skills must present themselves in a way that fosters understanding and applicability at the organizational level. By introducing and accepting these skills as competencies, health care organizations achieve a teachable model for their employees and clientele [53].The above section indicates the necessity, function, and benefits of treating virtue as a practical competency in health care organizations. By identifying virtue and understanding virtue as teachable and measurable competencies, health care professionals and health care organizations achieve analytic moral reasoning skills that aid facilitation efforts [54]. However, implementing analytic moral reasoning as a tool that aids conflict resolution efforts is best suited if these skills integrate into organizational policies. By first addressing how virtuous competencies integrate into organizational policies, methods of conflict resolution, ethical facilitation, and moral cooperation efforts become apparent. These competencies subsequently aid in bolstering the ethical conduct of a health care organization by introducing and teaching analytic moral reasoning skills [55].

An underlying goal of this analysis is to create a relationship between professional ethicists and health care organizations. This relationship is necessary for establishing hospitals as conglomerate moral agents and their accompanying health care teams. In effect, hospitals become moral agents themselves and function with inherent moral obligations within their practice [56]. This relationship is possible if individuals adopt virtue competencies to identify virtue in health care settings. Identifying virtue promotes analytic moral reasoning skills and subsequently bolsters the ethical character of an organization. However, introducing virtue competencies and analytic moral reasoning skills becomes a far more tenable feat if these competencies and skills assimilate into organizational policies [57].

Establishing a formidable ethical policy throughout a health care organization requires moral agents that both understand the moral goals of their institution and the processes necessary for its ethical success. Introducing the concept of moral agency throughout a health care organization is difficult to disseminate [58]. Moral theories and ethical principles are distinctive training competencies for health care individuals and typically aim toward individuals who specifically deal with moral facets of an organization, *i.e.,* legal departments, compliance, and ethics. However, introducing ascertainable moral agency and autonomy concepts creates an organizational attitude that is susceptible to learning analytic moral reasoning skills [59]. Although primarily directed for clinical ethics consultants, these skills and competencies can quickly assimilate into various positions throughout the organizational hierarchy. Furthermore, establishing an organization-wide understanding of moral agency aids conflict resolution efforts and bolsters the effectiveness of cooperation [60].

Due to healthcare organizations' hierarchical structure, the individual moral agency becomes efficiently introduced to an entire system if managers, staff leaders, and other higher officials are first informed and shaped into quintessential moral agents [61]. The subsequent watershed of moral fortitude increases if upper management understands the intricacies and effectiveness of developing moral agencies. The ethical goals of a health care organization should attempt to refine autonomous behaviors in a manner that promotes virtuous decision-making [62]. In this respect, leadership becomes a paramount concern for an organizational hierarchy. The sought-after qualities of an organizational leader involve various facets of ethical decision-making, stewardship, organizational integrity, and tactical planning.

Nevertheless, the primary obligation of a leadership role in a health care organization directs its duties toward cultivating autonomous behaviors that exercise ethical practice [63].Organizational ethics places a tremendous emphasis on employee decision-making. Whether an employee's duties involve micro-surgery or sanitation throughout the hospital, every employee has an opportunity to exercise their autonomy. However, it is up to the moral agent to determine what autonomous choices will foster a flourishing environment that promotes healing and well-being [64].

While ancient philosophical conceptions of virtue pertain to the betterment of the human soul, virtuous actions that apply to contemporary health care organizations focus on enhancing the effectiveness of appropriate medical and business practices. The betterment of the moral agent is still an effect of virtuous practice. Condemnation for obstructing virtuous policies in a health care organization possesses a two-fold consequence. First, deficiencies and excess of virtue, misidentifications of virtue, and a lack of analytic moral reasoning jeopardize the integrity and safety of the institution and its patrons. Second, the moral fortitude and virtuous abilities of the moral agent that engages in wrongdoing diminishes their occupational integrity and becomes susceptible to future wrongdoing, *i.e.,* medical error [65]. Since the results of improper analytic moral reasoning and virtuous practices are so detrimental to the mission of a health care institution, an organization's job must entail disciplinary actions. Often, conflict resolution involves reprimanding a blame-worthy party. Although unpleasant, this action itself exercises justice and attempts to reconcile organizational compliance procedures [66].

A moral agent should possess the capacity to decide what ought to occur in any given situation. Concerning organizational ethics, moral agents develop subsequent actions that translate a decision into a behavior. Leadership in health care organizations takes on a challenging role due to the responsibilities associated with upper management [67]. To reflect the responsibilities of moral agents in health care organizations, consider the following case: After a merging partnership with a subsidiary health care organization, the newly amalgamated billing department manager is seeking to fill a vacancy. The manager finds a seemingly perfect applicant who fits all professional criteria and demonstrates the necessary skills needed for the position. Later, the human resource office discovers that the applicant falsified travel reimbursement funds from the previous healthcare organization before the merge [68].

The above case pertains to ethical discrepancies that range from improper uses of autonomous decision-making, fraud, truth-telling—or lack thereof—and misappropriation, to name only a few. The case attempts to demonstrate the moral agency of the manager looking for a new employee and the moral agency of an ethics consultant responsible for assessing the facts of this case. While moral agency and autonomy are present within each stakeholder, this illustration concerns the application of virtuous moral agency and how autonomy applies appropriately. This application bolsters the educational possibilities of ethics and influences the normativity of ethical deliberation at the organizational level [69]. The manager is in a position of leadership that requires advanced skills in analytic moral reasoning. Without these skills, the outcome of the above case has an increased probability of ending detrimentally. The necessary skills that dictate virtuous leadership techniques aid in conflict resolution and facilitate ethical discrepancies at the clinical and organizational levels. With the analysis of the moral agent illustrated, the following section presents distinctions in moral theory and presents the ethical qualities of a virtuous leader in health care organizations [70].

Identifying virtue in health care situations, both clinically and organizationally, certainly aids in resolving ethical discrepancies. However, health care organizations benefit tremendously by implementing virtuous tactics in their deliberative processes by reconciling business conflicts, hierarchical discrepancies, and policy development [71]. This analysis demonstrates the importance of moral agency and autonomy, yet applying autonomous behaviors that align with virtue benefits a health care organization when the moral agent is in a leadership position. Examining the qualities of a morally virtuous leader indicates further benefits of ethical practice in organizational ethics and professional health care [72].

Ethical decision-making at the organizational level in health care is a complex and challenging endeavor. This difficulty arises primarily from the consequences that arise from organizational decisions and the scope of parties affected. Resolving a quarrel between two involved party members is comparatively more straightforward than resolving a dilemma for an entire enterprise [73]. However, while the implications and repercussions from organizational decision-making are tremendously impactful, the basis upon which these decisions ground themselves involve the same ethical decision-making processes as prototypically minute ethical situations. By identifying and exercising moral virtue at the corporate level, analytic moral reasoning skills foster formidable attitudes of autonomous leaders [74].

Improving methods of conflict resolution and ethical facilitation require successful leadership qualities at the organizational level. Virtuous leaders possess the ability to recognize unethical practices and instances where agents improperly implement degradations of virtue. Organizational leadership is a critical component to a virtuous health care organization because virtuous leadership shapes and perpetuates how autonomous decisions become aligned with an organization's mission of stewardship and integrity [75]. Besides the ability to recognize excesses and deficiencies of virtue in a health care setting, the qualities of a virtuous leader cultivate an environment that promotes collaborative efforts. These efforts aid in conjuring an environment where all employees, lay-persons and professionals alike, desire to contribute to the overall mission and purpose of the health care organization. Virtuous leadership is attentive to employee's needs and empowers individuals to become self-motivating. Additionally, virtuous leadership contributes to conflict-resolution efforts and moral cooperation standards by establishing a communicative buttress that weds and supports ethical practice's private and public realms [76].

Virtuous leadership in health care is not strictly reserved for corporate figures and administrative positions in upper management. The beauty and effectiveness of analytic moral reasoning skills derived from identifying virtue lie in the plasticity and applicability of leadership throughout an institution. This analysis emphasizes courage and justice as illustrative virtues in a health care institution. However, atypical virtues that apply to health care organizations also serve as formidable. Concepts like attentiveness, empathy, healing, moral persuasion, foresight, and concept analysis are acceptable virtue-competencies that, like courage and justice, are capable of deficiency or excess [77]. Virtuous leadership in a healthcare organization promotes identification abilities and analytic moral reasoning skills derived from this identification. For instance, an excess of attentiveness may result in over-analyzing a situation and thus establish an environment that does not allow others to perform their tasks independently or effectively. Alternatively, a deficiency of inattentiveness may disregard important information, leading to patient endangerment or medical error [78].

Virtue identification and subsequent analytic moral reasoning skills have tremendous benefits at the organizational level in health care. These skills permeate and apply across an entire system if appropriately enacted. However, the acceptance and practice of these skills take another effect if the adoption of these skills possesses organizational repercussions if not followed accordingly [79]. Synonymous with their accompanying practical skills, *i.e.,* the ASBH core competency skills for clinical ethics consultation, the adoption of analytic moral reasoning skills must assume a critical role and demonstrate substantial influence. To reinforce the practical identification of virtue and the resulting skills that follow, health care organizations must determine if blame-worthiness and disciplinary sanctions are necessary components for ethical success [80].

Fostering a health care environment that promotes safety and stewardship requires an organization that prioritizes institutional policies that thwart detrimental outcomes. One of the most formidable and long-standing traditions in bioethical facilitation includes the principle of cooperation. A historically Catholic concept, the principle of cooperation is far easier to understand than apply [81]. Moral cooperation bases itself upon intention. If a moral agent intends to cooperate with an immoral act, he is morally culpable. If a moral agent unavoidably or unknowingly associates or involves themselves with a morally wrong action of another, he is not morally culpable. It is essential to note the distinction of intention when addressing moral cooperation. If the moral agent does not intend to perform an immoral act, he is not morally culpable. For example, if some individual loads a handgun and gives the firearm to another individual under the firm impression that the recipient intends to shoot a target, but the recipient shoots an innocent bystander with malicious intent, the individual who loaded and handed the firearm to the recipient is not morally culpable because their intention was not to commit or aid in the immoral act. However, if the individual loading the firearm hands it to the other individual, fully understanding that the recipient uses the firearm to commit a malicious or evil act, they are morally culpable. The moral culpability of the agent assumes the agent's intent to cooperate with a given act [82].

The above description and illustration of moral cooperation are applicable within health care organizations. A hospital administrator who violates corporate policy under pretenses due to another individual's influence does not willingly cooperate with the act in question. Alternatively, the same administrator who signs a document that intentionally violates patient rights for monetary gain is morally culpable, despite never directly interacting with patient finances. Moral cooperation plays a vital role in conflict resolution due to its ability to discern intention and involvement in immoral acts [83]. Moral cooperation becomes complicated when attempting to discern an individual's involvement with an act that violates organizational policy. Proving illicit formal cooperation or the cooperator's direct consent to partake in an evil act by directly assisting in executing the act becomes problematic when multiple party members and medical facts are involved. Even more onerous is the process of justifying and identifying licit material cooperation or reluctant cooperation with an immoral act that another individual performs [84].

The excesses and deficiencies of virtue are inherently involved in illicit formal cooperation –the form of blame-worthy cooperation and require disciplinary actions [85]. For instance, the organizational administrator who willingly signs a document to undermine patient rights for monetary gain demonstrates a deficiency of integrity and stewardship. Alternatively, actions that exercise a proper balance of virtue reveal themselves in individuals who cooperate in an immoral act but have no intention of committing the immoral act, *i.e.,* a surgical nurse who intends to do nothing more but prepare a well-organized and sterile operating room for a surgery that terminates a viable pregnancy in a catholic hospital. The surgical nurse has no intention or knowledge of a prohibited abortion or violates hospital policy. The surgical nurse exercises virtuous behavior by performing his task well and engaging in his duties under the assumption that the surgical team will perform their task in compliance with organizational policies. Although the abortion could not have taken place without a ready operating room, the surgical nurse did not know nor intend to engage in the illicit act [86].

The above examples illustrate how moral cooperation manifests and the usefulness of analytic moral reasoning skills—both scenarios exhibit opportunities for virtuous recognition. Whether the recognized instances of virtue entail deficiencies or excesses, the mere act of recognizing virtue-related instances promotes analytic moral reasoning skills for moral agents [87]. However, it is essential to note the differences in educational opportunities. While the surgical nurse is already engaged in his job virtuously, one must assume that he possesses analytic moral reasoning skills. The surgical nurse has no reason to believe the instruments he prepares to serve as means to an evil end. However, the physicians performing the procedure and the hospital administrator in the previous example face an opportunity to exercise analytic moral reasoning skills. Understanding the implications and intentions of action are learning experiences if a health care organization recognizes these opportunities and capitalizes on these educational instances. By implementing virtue ethics into a structured program in teaching hospitals, fellowship programs, and internships, health care organizations can foster morally fortuitous attitudes in health care [88].

The cornerstone of this analysis demonstrates the methods and practicality of introducing virtue-identification in health care organizations. The benefit of honing virtue-identification techniques lies in the subsequent development of analytical moral reasoning skills. These skills, although tenable, still require a suitable medium that teaches and reinforces these skills. While the practical skills in the ASBH's core competencies for clinical healthcare consultants are composed in a teachable format, these same skills are infinitely more accessible and effective if analytic moral reasoning skills are assumed to be a prerequisite condition for clinical consults [89]. Despite the various existing preliminary programs in ethics education, *i.e.,* fellowship programs in ethics, internships, etc., these programs do not reinforce the techniques and educational facets illustrated in this analysis. Aiding current efforts to bolster ethical practice in health care organizations and professional ethics is deficient in many respects. Thus, the following section demonstrates the effectiveness of refined organizational curriculums for medical residents, ethicists, administrators, and other involved stakeholders. In doing so, these integral methods demonstrate the moral prowess of health care professionals and the inherent virtue embedded within clinical ethics consultation careers [90].

The most commonly recognized health care preparatory techniques are medical residencies. These programs, depending on specialty, vary in duration, educational information, and procedural experience. Residencies are designated only to health care organizations that can accommodate a teaching staff, curriculum, and venue for recent medical school graduates. While the structure and variance of medical residency programs are key factors in curriculum design, programs relate to one another with the intention of formal professionalization [91]. Similarly, health care professions like nursing, ultrasound-technicians, and nutritionists possess a formal introduction to a news organization. Often, the training for these professions manifests as internships, probationary learning periods, or residency programs. Although the curriculum's variations depend upon organizational policy, each specialty usually accompanies three significant educational facets: Classroom work, practical or hands-on experience, and professional review [92]. While these health care curriculums are supported and regulated by a governing body, *i.e.*, ANCC for nursing, ABPS for physicians, etc., clinical ethicists lack a proper certifying administration. Students receive clinical ethics education module training throughout their programs, but not all programs give students a genuine clinical experience. Students receive necessary information in didactic sessions, including the normative ethical theories that support these modules. However, these ethical training modules primarily manifest in didactic sessions and not in a clinical setting [93]. Clinical ethics is a vocational discipline that requires training in normative ethics and practical application. As such, clinical ethicists must engage with a clinical venue and the stakeholders within it.

The stringent requirements health care organizations coincide with one another and cooperate with the organizational standards of a corresponding certifying body. While no such institution exists for clinical ethics consultants, organizations like the ASBH and the CECA subcommittee of the ASBH are working diligently to establish certification criteria [94]. However, the classroom work, practical or hands-on experience, and professional review accompanying residency programs can apply to residency programs in a seamless fashion. Synonymous with organizational leadership, a health care organization that serves as a venue for residency programs, internships, and fellowships alike must equip themselves with ethical criteria that support the institution's mission. Additionally, a health care organization that intends to integrate a program that teaches, certifies, and bolsters ethical education must implement the virtuous identification tactics illustrated in this discussion. Implementing these tactics is pivotal for fostering analytic moral reasoning skills amongst ethicists and health care professionals alike due to the lasting and applicable quality of moral reasoning skills across a health care institution [95].

Health care organizations' variability can accommodate a virtue ethics curriculum by adopting simple organizational principles. These principles are synonymous with the qualities associated with ethical leadership illustrated earlier in this analysis. However, implementing virtue ethics into organizational curriculums must conform to a model that aids in shaping moral character across an institution [96]. Before considering a virtue ethics curriculum, the institution must possess a code of ethics that does not conflict with other facets of organizational policy—additionally, articulating the institution's code of ethics to employees in a way that is accessible and unambiguous. The organization must have an existing protocol for resolving conflicts and grievances. Finally, the prerequisite criteria must possess a protocol for enforcing their code of ethics in a reasonable and just manner [97].

Accreditation for hospital instantiation must possess the facets mentioned above to gain approval and license to practice under an appropriate governing body, *i.e.*, the joint commission of hospital accreditation. However, due to the current lack of a formal accreditation body for assessing ethically relevant education modules, a virtue ethics approach to organizational and professional ethics is a tenable alternative that can integrate into current organizations and become a lasting model amidst future accreditation efforts [98]. Introducing virtue ethics techniques into residency programs, fellowships, and other forms of professional education is possible, provided the existing organization possesses the ethical criteria mentioned above. However, the institutionalization of virtue ethics is a process that requires a professionally trained ethicist to relay information to health care professionals and other budding ethicists [99].

Finally, establishing an organization that promotes and teaches virtue ethics to its residents and other associated staff promotes an ethical culture that promotes virtuous behavior across an entire system. A health care organization that teaches virtue identification techniques yields a culture of analytic moral reasoning [100]. These skills allow professionals to delineate professional roles and boundaries, facilitate ethical decision-making, resolve medical and organizational conflicts, and contribute to a widely shared philosophical understanding amongst involved organizational stakeholders [101].

To institutionalize virtue ethics throughout health care organizations, a series of professional suggestions expedite the integration of newly-established teaching methods. First, under the guise of virtue ethics education, organizations develop formidable techniques that delineate professional roles. Understanding the differences in professional roles bolster progress and limit conflicts related to responsibility and duty. Second, an organization trained in identifying virtue and subsequently develops analytic moral reasoning skills dwell in an advantageous position to develop a code of ethics that expands upon an existing code of ethics and refines an organization's purpose and mission [102].Under the guise of virtue ethics, the newly established mission of a health care organization promotes desired ethical behaviors and aids in addressing "grey areas" incorporate clinical and social decision-making. Finally, establishing an organizational culture that fosters virtue ethics inevitably redefines a corporate ethics committee that curtails education, policies, and regulatory functions for professionals across a system [103].

The implementation of virtue ethics throughout a health care organization is not only possible but necessary. Virtue serves as a positive addition to an organization's framework due to its ability to instantiate moral reasoning skills that promote an ethically-oriented culture permanently. Most importantly, facilitating and implementing virtue ethics throughout a health care organization eliminates unethical behaviors and bolsters the productivity and effectiveness of a healing enterprise [104].

While this discussion attempts to demonstrate the lasting utility and efficacy of virtue ethics and analytical moral reasoning skills throughout health care organizations, virtue ethics also indicates the nature and scope of clinical ethics consultation. Clinical ethics consultants are inherently involved with an occupation that involves the integration, implementation, and dissemination of virtue [105]. While virtue manifests in different instances, including its deficiencies and excesses, it is the task of the clinical ethicist to recognize and explain how virtue manifests. In this respect, the trained ethicist is the first line of defense for ethical deliberation. From this standpoint, the clinical ethicist is responsible for not only assessing instances of virtue, its deficiencies, and its excesses, but he is also responsible for articulating the methods in which virtue articulates in a health care setting. Furthermore, since the ethicist is familiar with the ethical concepts derived from virtue, he is also responsible and obligated to educate other professionals in health care about virtuous practices and how to regulate their application [106].

Clinical ethics consultants cannot perform their tasks effectively without a firm basis in philosophical concepts. However, due to the multidisciplinary aspect of health care ethics, comprehensive education in the history and moral theory is not a practical endeavor, nor is this approach beneficial for health care ethics [107]. While the multidisciplinary aspect of health care ethics consultation does not comprehensively partake in an education rooted in the history of philosophy, individuals who choose to pursue a career in health care ethics can bypass a formal education in philosophy by developing virtue-identification skills. Consequently, these individuals from different backgrounds attain a philosophical grounding in moral theory by obtaining analytical moral reasoning skills [108].

Analytical moral reasoning skills allow individuals from various disciplines to take a philosophical approach to health care situations without extensive experience with the philosophical, moral corpus. In this respect, clinical ethics consultants join a recognized profession. However, it is essential to distinguish the profession's virtuous nature and its individuals [109]. While the nature of the profession itself is virtuous because clinical ethics consultations—under the guise of analytical moral reasoning skills—require identifying and implementing virtue, the individuals engaged in the profession do not possess inherent virtue. While this distinction does not attempt to ascribe moral priority to consultants, this discussion does assert the virtuous nature of clinical ethics consultations [110].

Perhaps the most significant aspect regarding the virtuous nature of clinical ethics consultation involves the dissemination and growth of virtuous, ethical cultures within health care organizations. Organizational culture shares principles a like-minded group accepts to solve problems and integrate policies that reconcile issues and prevent future discrepancies [111]. Due to clinical ethics consultation's ability to promote virtuous activity throughout an organization, clinical ethics consultants also promote an atmosphere of the organizational moral agency. Although moral agency prototypically pertains to individual autonomy, and practical clinical ethics consultant team can foster an adaptive moral learning atmosphere to educate an entire health care organization. The power and influence of a virtuous profession like clinical ethics consultation revitalize learning strategies, organizational rationality and aids in preventing organizational failures. Consequently, health care ethics consultation is inherently a virtuous profession due to its promotion and dissemination of virtuous behaviors [112].

While the organizational facets mentioned above demonstrate the difficulties associated with integrating virtue ethics into a wide-span consultation curriculum, the possibility of expanding an ethics consultation curriculum to ethics consultants and medical professionals is also a tenable endeavor. Moreover, utilizing residency programs as a venue that supports and disseminates necessary information to clinical ethicists in training has proven to work well in an educational fashion. The ASBH core competencies for clinical ethics consultants have laid significant groundwork for a more developed consultation curriculum in the United States.

Combining the ASBH's practical knowledge and the analytic moral reasoning skills acquired from virtue identification mentioned in this analysis, a comprehensive ethics education is bolstered and has a significant stake in professional health care.

Although this analysis thus far has advocated for implementing additional cur-riculum points to solidify a comprehensive professional ethics program for consul-tants, there inherently lie several issues that accompany this endeavor. This analysis does not intend to compose and disseminate an ethical lessonplan to budding ethi-cists and health care professionals. Instead, this analysis proposes virtue elements of moral reasoning that should accompany a professional credentialing program for clinical ethicists. In doing so, this tactic may bolster an ethicist's understanding of clinical ethics and function in a manner that emphasizes analytic moral reasoning skills in practical health care situations. By examining applied program issues, this analysis does not intend to solely identify the issues that exclusively accompany an ethics program for clinical ethicists and examine the difficulties associated with implementing virtue ethics curriculum points into a practical education model, i.e., into a practical education model a venue like a residency program. This process is elucidated by examining how new knowledge points integrate into a new curricu-lum and how existing programs have introduced novel education points.

The breadth of literature about ethics education is vast and yet simultaneously lacking in substantive content [113]. While various texts attempt to inform budding ethicists about the methods and tactics used in practical facets of clinical bioethics, there is a significant gap in the corpus regarding philosophical contributions about contemporary methods of bioethics consultation [114]. Furthermore, there is sig-nificant literature on the effectiveness of virtue in medicine and lacks literature about the effectiveness of understanding virtue in ethics consultation. In this respect, the contributions presented in this analysis aid the development of clinical ethics consultants by introducing an accepted methodology to a developing professional discipline [115].

A quintessential ethics training program description elucidates how virtue train-ing can effectively develop certification programs for clinical ethicists. Furthermore, a description of ideal ethics training for clinical ethicists, under the guise of this analysis' key components for an ethics consultation curriculum, unearths subse-quent issues that require attention. However, the cultural roadblocks in hospital medicine that prevent ethics from thriving as a professional discipline should receive attention since these barriers affect how a quintessential ethics program may mani-fest. Assessing what cultural barriers stand in the way of clinical ethics aid this analysis by demonstrating what measures may encourage the acceptance of clinical ethics in professional health care. The following subsections describe the difficul-ties associated with clinical ethics consultants' current cultural climate and their programs. Additionally, the following subsections address the difficulties associated with ethics access in contemporary hospital settings [116].

To ascertain how an ideal ethics program functions, the health care community must view clinical ethics consultation as a service that is both accessible and effec-tive. If viewed as a regular consultation service, ethics consultation services should function like any other medical consultation specialty, i.e., gastroenterology, cardi-ology, etc. In contemporary medical practice, specifically in American hospital sys-tems, a patient is under the medical care of many individuals whom an attending physician leads. Although responsible for a patient's care, this attending physician is a specialist in a specific area of medicine. While many attending physicians specialize in specific fields, many attending physicians are trained explicitly as hospitalists; skilled in-patient physicians who work exclusively in hospital settings [117]. Although patients or their surrogates are responsible for expressing values, wishes, and appropriate medical treatments, it is the responsibility of the attending physician to decide whether to abide by these wishes. Additionally, the attending physician has the authority to request additional consultations, place medication orders and codes, and ultimately determine the course of treatment for a patent by acquiring and implementing relevant information [118].

The role of the attending physician is pivotal and possesses a fair amount of weight in clinical decisions. An attending physician ultimately requests this service to obtain this additional resource. However, while the attending physician has the final word in ordering an external consultation, resident physicians can request ethics consultations, other consultants involved in a patient's case, and even patients themselves if they express a medically relevant concern about their care [119].

Additionally, ethics consultations are far more effective if a hospital's culture views an ethics consultation service as an individual entity consulted like any other medical specialty [120]. While the ethics consultant's expertise determines if a consult is a relevant ethics consult, it is the prerogative of all individuals involved in a health care system to request an ethics consultation [121].

Despite ethics consultations' inherent nature and availability, ethics consultation services are often disregarded, feared, or simply unknown [122].In various instances, physicians condemn, are often discarded by attending physicians due to several issues. This section discusses three primary reasons why cultural attitudes in health care disregard or do not utilize ethics consultation services. First, various attending physicians do not understand the purpose and scope of clinical ethics. In this respect, many physicians believe ethics consultants investigate gross misconduct, enforce compliance procedures, and implement disciplinary actions. These facets of professional investigation are not within the purview of clinical ethicists. Instead, these facets belong to human resource representatives, compliance officers, and legal departments. While these departments investigate gross misconduct and determine disciplinary actions, ethicists can still assist if there is an ethical discrepancy that demands the expertise of a moral agent [123]. By addressing these issues in a manner that demonstrates the scope and practice of clinical ethics, physicians may view ethics consultation services as a resource rather than a threat. Although various interventions may accomplish this task, this analysis contends that the presented critical components for an ethics consultation curriculum ultimately aids the eventual acceptance of professional ethics consultations in health care.

A second conception of clinical ethics that deters physicians involves the chain of command in hospital settings. Many physicians are often not accepting of consultations that have not first gone through their approval. While a patient's attending physician has the final approval for written orders, ethics consultations are unique because all stakeholders can request ethics consultants. This issue causes duress for attending physicians, mainly due to clinical ethicists' interactions with patients, their families, surrogate decision-makers, and other medical professionals [124]. If viewed as a service, patients, team members, and other stakeholders receive professional advice from neutral individuals in their approach to conflict and thus quintessential mediators during value-laden uncertainty.

A final reason physicians' reluctance to effectively implement the assistance of ethics consultation services lies in availability issues. While many hospitals require an ethics presence for accreditation, only a select few facilities possess the resources for an individual-based consultation model for clinical ethics consultation [125]. Most hospital-based ethics programs reside in teaching facilities, *i.e.,* Mayo Clinic, Wellstar Health System, Cleveland Clinic, UCLA, etc. These facilities can accommodate an individual-based consultation model for clinical ethics consultations due to their resources. Additionally, these facilities have a propensity for various types of education in health care due to their status as medical education institutions [126].

If the greater medical community does not recognize clinical ethics as a professional discipline, the development of ethics programs slows tremendously and delays the task of solidifying ethics consultation service as a professionally recognized discipline in health care. Consequently, the proposed curriculum points in this analysis do not have a venue to manifest. However, although they may not become resolved in this analysis, these issues should receive attention. While it is difficult to resolve or change the cultural dichotomy of American health care, it is undoubtedly not an impossible endeavor. By merely identifying the primary barriers ethics consultants face, the possibility of establishing a consultation curriculum in an environment that facilitates all facets of health care becomes far more tenable. While this analysis intends to identify and introduce key components for an ethics consultation curriculum, the larger picture of this endeavor is to aid the effort in establishing ethics as a normative and professional discipline [127].

A description of a quintessential ethics program in a hospital setting responds to these issues and demonstrates how the key curriculum points in this analysis are applied to address the issues that thwart the cultural acceptance of clinical ethics involvement in health care. Additionally, this contribution aids in developing a quintessential consultation education program for clinical ethicists. Although not the focal point of this project, it is helpful to examine how a clinical ethics consultation curriculum that implements the virtuous key curriculum components mentioned in this analysis may manifest in an educational setting. While this described concept of an ideal clinical ethics program intends to maintain the virtue components mentioned in this analysis, ethics departments may remain malleable in their development and overall function in an institution. The malleability of a department allows room for change and, like the founding fathers of American bioethics, pragmatic development [128].

Clinical ethics departments ought to function with a sole-consultant model. Instead of referring specific cases to a board of ethics committee members, involved stakeholders should access a 24/7 ethics service to attend to ethical discrepancies in real-time [129]. These consultants work on an on-call schedule to accommodate the needs of a hospital system regardless of size or operational purview.

## Training Programs for Clinical Ethicists

Residencies and fellowship programs serve as appropriate venues for implementing this analysis's model of education. Ethics consultation programs are structured around an institution's guidelines and are[typically] contained within an existing educational venue, *i.e.,* a teaching hospital, in full compliance with their corresponding accrediting body [130]. These accrediting bodies include the Accreditation Council for Graduate Medical Education (ACGME), The Joint Commission, etc. [131] Hospital residency programs for clinical ethicists work and perform their residency alongside medical residents, hospital chaplains, social workers, nurses, and all other specialties that possess residency programs and require ethical facilitation within their practice [132].A clinical ethics training program seeks towed vocational ethics with health care systems, specialists, professionals, and departments [133].Each training program must receive accreditation by a corresponding accreditation body deemed appropriate by each teaching hospital's administration, *i.e.,* Clinical Pastoral Education program (CPE) certified through the ACPE, NACC, or other appropriate accreditation body for given healthcare-associated occupation [134].

Ethicists in residencies should perform daily rounds, document, keep track of patients and attend regular didactic courses [135]. These courses combine a continuing education curriculum with specific clinical instances within a health care setting. Ideally, ethics residents will review the same patients treated by medical residents, social workers, nurses, and other newly inducted residents [136]. Furthermore, residency programs should work together as much as possible [137]. Ethics residents and medical residents may attend the same post-round meetings to learn from each other's observations, reflections, and clinical expertise. In effect, medical residents better understand ethical conduct in their practice, while ethics residents better understand the medical jargon, clinical situations, and patient treatment options [138].

While the duration of each residency program becomes established by the teaching hospital, accrediting body, or both, residency programs for ethicists should take 1 to 2 years [139]. Each program is largely self-governed under the knowledge bases outlined in this assessment. However, a large portion of residency training for clinical ethicists involves applying moral philosophy with the situational aspects of health care [140]. Naturally, ethicists cannot perform this task without the skillsets presented by the ASBH. By emphasizing both aspects of the residency curriculum, ethicists become prepared to address various issues throughout a health care institution [141].

Coupled with the criteria set out in the CECA report, a requirement of clinical ethics residents involves keeping a log of clinical cases. Implementing a documented portfolio that maintains patient confidentiality may become an effective tool in demonstrating the effectiveness of consultation and the skills and knowledge possessed by the resident [142]. Recorded mock consultations should become utilized to evaluate individual residents. A requirement of ethics residency involves peer and administrative review of these portfolios to complete the residency program successfully [143]. Additionally, regular examinations keep track of each ethics resident's progress. The residency program concludes with a practicum examination as well as a final written examination [144]. These exams intend to assess the resident's theoretical knowledge competency regarding the moral theory and the standards and practices of contemporary health care ethics consultation. After a complete residency program, passing these exams ensures licensure and accreditation, provided the ethics resident is in good standing with their coursework in the residency program [145]. The success of the proposed vital curriculum components hinges upon implementing a sole-consultant model and is taught similarly in content and form as the program described above.

Contrary to ethics committee models, a sole-consultant model is a far more effective way to deliver practical ethics consultation and disseminate ethics education throughout a hospital system. Moreover, a sole-consultant model is a far more manageable venue in which virtue ethics components may manifest. While ethics committees possess an essential function and role within hospital ethics, expecting a team of health care professionals to dedicate their time to understanding and implementing moral virtue in their practice is too onerous [146]. Alternatively, a sole-consultant model operates where individual consultants—trained in moral theory—handle cases personally and subsequently deliver information to the ethics committees for educational purposes. While the ethics committee under a sole-consultant model is not responsible for clinical ethics consultations, the committee plays a role in quality improvement work, education dissemination, and community outreach.

Additionally, ethics committees are remotely involved with consults under this model when consultants require a hospital community's input and moral determinations [147]. Since hospital cultures and communities vary, it is wise to gather the perspectives and attitudes of individuals who compose a hospital ethics committee. This tactic derives cultural attitudes of a health care community and aids a consultant's ability to deliver practical recommendations [148].

The sole-consultant model is effective and remedies the primary issues associated with clinical ethics involvement in hospital medicine. Specifically, implementing a sole-consultant model accomplishes focal tasks: First, the sole-consultant model effectively introduces ethics into a health-care culture that aids in eliminating pre-conceived notions of ethics. These pre-conceived notions include viewing ethics as a disciplinary department, a department that investigates gross misconduct, and a department that delivers legal advice [149]. The sole-consultant model alleviates these misconceptions by placing individual consultants at the bedside with attending physicians and other medical professionals. Second, in conjunction with modified ethics committee functions, the sole-consultant model provides effective information delivery methods [150]. The lack of ethical awareness in hospital settings is often rooted in unavailable services. Existing ethics consultation services are hospital ethics committees. These committees comprise individuals who often have minimal ethics training and come from a diverse background of expertise. By implementing a sole-consultant model, committees can provide additional functions like quality-improvement projects, community outreach efforts, and hospital-wide education. By disseminating information, medical professionals subsequently learn the proper function of clinical ethics and are thus more likely to accept these services. These methods alleviate the burden of other departments and work toward establishing an ethically rich culture throughout a health care institution [151].

This analysis does not suggest that ethics committees are ineffective in their tasks. On the contrary, the sole-consultant model for clinical ethics does not work without an ethics committee that performs its functions well. The key curriculum components in this analysis offer skills in analytic moral reasoning that apply to both clinical situations and organizational leadership. A consultant trained under the same criteria outlined by the ASBH core competencies in conjunction with the key curriculum components of virtue identification yields a professional who can function as an effective consultant for value-laden discrepancies in health care. Additionally, an appropriately trained clinical ethicist can guide and train others interested in clinical ethics. While this model remedies the issues mentioned above, additional issues lie in the organizational sphere of health care, including policy development, quality improvement, and root cause analyses [152].

The utility of this analysis has manifested in identifying essential facets of a clinical ethicist's curriculum. However, the critical components for an ethics consultation curriculum reside in applying and implementing virtue identification for clinical ethicists in health care settings [153]. While these skills are applicable in public and private life, the utility of virtue identification in work associated with clinical ethics consultations aid a consultant's ability to morally deliberate and thus present formidable recommendations for affected stakeholders in ethically precarious situations. Nevertheless, teaching virtue becomes an onerous task if an existing curriculum does not implement virtue in its lessons. Additionally, teaching virtue is difficult if there is no formal method for relaying otherwise abstract information. Thankfully, Aristotle's presentation of virtue in his philosophy of ethics assures that virtue is a teachable skill that can be refined and developed with practice [154].

Explaining how virtue may be taught to clinical ethicists ultimately demonstrates how the critical components for an ethics consultation curriculum manifest. The previous components of this discussion all include skills and knowledge points that, presumably, can be taught and refined through their implementation and use in clinical ethics [155]. For example, using the formidable consultation methods elaborated in chapter three can become more effective if rehearsed and implemented in clinical cases. Nevertheless, virtue seems to allude to this development due to its abstract nature and theoretical framework. In his *Nichomachean Ethics*, Aristotle asserts that virtue can indeed become more refined with practice. Like any other practical skill such as carpentry, sailing, or mathematics, virtue can be taught and refined through repetition and hermeneutic development [156].

While this rather abstract assessment of virtue identification and practice is clear, these methodologies and theories still need a practical application into a formal certification program. In this respect, how clinical ethicists may become versed in virtue ethics and virtue identification can manifest in a somewhat objective format, along with the various educational components that clinical ethicists need to know. A sound curriculum can teach virtue identification techniques to collaborate with other educational facets of a clinical ethics consultation. First, education techniques for teaching virtue identification may manifest in practical instances of clinical ethics. Clinical ethicists in training may receive various mock cases where the consultant must identify the central ethics issue. In doing so, the clinician may adjudicate between possible actions of stakeholders involved and how these actions weigh against excesses and deficiencies of virtue [157].Once the consultant identifies these extremes, the consultant may then discern the mean between these two extremes, thus identifying the virtuous action in the given clinical situation. Subsequently, the clinician may act accordingly by implementing recommendations that coincide with the identified virtue. By way of example, consider case 4.1 in the *cases* section of this text.

Properly trained clinical ethicists should possess the necessary knowledge points to address a case like 4.1. In terms of practical virtues, Aristotle notes that courage comes before all other virtues, and all other virtues rely upon courage [158].Courage is easily identified as a focal virtue that ethics consultants may use to discern proper ethical recommendations for the care team. However, merely identifying courage as a virtue associated with this clinical case simply because Aristotle claims that courage is at the root of all virtues is not enough to justify the identification of virtue in this instance [159].

Clinical ethicists must identify virtue by identifying the excesses and deficiencies of the virtue in question. Regarding case 4.1, various virtues in this scenario could receive credit. Virtues like courage, justice, temperance, and poise are virtues that one may justify in this clinical circumstance [160].To test and hone a clinical ethicist's abilities in identifying virtue and virtue's absence in clinical situations, clinical ethicists must indicate excesses and deficiencies in virtue within their vocational practice. An evaluation of virtue in clinical cases allows the proper development of recommendations for clinicians [161]. For example, case 4.1 displays various deficiencies in virtue that could manifest from specific paths that the clinical ethicist and the team may take. Continuing full medical efforts, including resuscitation, is undeniably violent and seemingly futile. Though the patient has the righto requesting the team's continuation of resuscitative efforts, care teams must do no harm and ration or triage their resources to treat many patients at once.

Developing algorithmic or otherwise value-based systems for rationing resources and preventing overall harm requires courageous action and endurance through a tumultuous clinical situation. The number of resources required for ongoing treatment and resuscitation is extensive. Continued resuscitative efforts for the patient in 4.1 require various resources like the time and effort of the care team, blood products, mechanical ventilation, ongoing ECMO, CRRT, crash cart accessibility, oxygen, and chemical resuscitation products. Naturally, a principlistic model comes to mind when assessing a case with high complexity. The justice aspect of resource allocation and fairness to the patient and his family is apparent, but the dichotomous relationship among beneficence, nonmaleficence, and autonomy is perhaps the more questionable facet of this case. The patient and his family are clear in their therapeutic pursuits, but adhering to a patient and family's request during a medically futile situation is daunting. While the medical team *can* continue the patient's existing therapies, the question of *should* is ambiguous.

Disregarding the foundations of principlism is an error in this case and clinical ethics consultation generally. Principlistic facets in a given ethics case should receive attention and serve as foundational identifiers of ethics issues. In conjunction with a principlistic determination, the virtue components in this analysis aid in delving deeper into the underlying ethics issue and aid in determining ethically supportable actions.

Often, patient and family requests that accompany significant harm are not an emphasized concern. Instead, families and patients seek a therapeutic resolution and believe the harm associated with their request is a small price to pay if extending an individual's life is the outcome. Often, how an individual exists is far less concerning than the individual's non-existence. Regardless, merely identifying the excesses and deficiencies in virtue in this case, thinking through these processes facilitates analytic moral reasoning and promotes analytical moral reasoning skills for clinical ethicists [162].

Virtue identification stands as the focal key curriculum component for an ethics consultation curriculum. Identifying virtue in various clinical situations aids a consultant's abilities by helping him tap into the knowledge and skillsets established through formal education techniques and allows the consultant to engage in ethics cases critically. This method also allows clinical ethics consultants to implement moral theories like utilitarianism and deontological methods. However, teaching this reasoning model requires various mock instances that test for and evaluate one's ability to identify the excesses and deficiencies of virtue in clinical situations. While this novel technique in shaping clinical ethicists possesses an array of challenging educational facets, it becomes far more effective if taught in a sole-consultant model of clinical ethics.

The sole-consultant model serves as an excellent venue to accommodate the proposed curriculum points mentioned in this analysis. The implementation of a sole-consultant model remedies various issues that are currently present in hospital-based clinical ethics. However, the scope and purview of clinical ethics are still significantly undermined when an ethicist's duties do not affect other medical professionals [163]. The role of a professional in a medical setting should directly impact patients, providers, and administrators. While clinical work significantly impacts a health care system, the effects are often not viewed on a system-wide scale. This is a central issue surrounding clinical ethics consultation and is yet another blockade to professionalize clinical ethicists [164].

Implementing the key curriculum components into a clinical ethicist's education is paramount for the field's professionalization. It is equally important that the enormous scope of medical professionals understand the utility of this discipline. By examining the scope of clinical ethics, the broader utility of the discipline becomes apparent. Specifically, examining the scope of clinical ethics outside of bedside consultants illuminates moral theorists' utility in organizational practice and quality [165].

# Ethics Consultation Departments and System Operations

An ideal ethics consultation department ought to function alongside four influential groups in health care. These groups include palliative medicine, social work, and chaplaincy. Naturally, these disciplines have their functions and do not comprise various departments where clinical ethics is applicable [166]. Ethics consultants aid the departments mentioned earlier by assembling data from each department's perspective [167]. Issues within a hospital system require specialized attention. While ethics consultation does not intend to infringe upon these areas, it is still within the scope of ethical practice to aid these areas of expertise when ethical discrepancies arise, particularly within patient care [168].

Quality departments typically assess patient safety, gaps in care and aim to improve the overall functionality of a health care system. For instance, a hospital's mortality metric may indicate that the monthly deaths at a specific institution are much higher than national data or neighboring healthcare systems. The quality department is responsible for tracking the mortality trends among other systems and comparing their data. At the same time, this task is a statistical endeavor, the actions within the quality department attempt to alleviate volatile practices in hospital medicine [169]. Quality departments use statistical methods to evaluate the cause of issues like increased mortality. Furthermore, these issues cannot resolve without an action plan. Clinical ethics consultants can have a tremendous impact on quality initiatives due to ethical work in hospitals.

Expanding upon this example, some hospitals assess mortality metrics with code status determinations upon admission. Patients who enter a hospital with a do not resuscitate order or change their code status to a do not resuscitate order within the first 24 hours of their admission do not impact a hospital's mortality metric upon the patient's demise [170]. While quality improvement departments identify this process, working with ethics departments can help curb a hospital's mortality metric by engaging in value discussions with patients upon admission. Patients typically provide their code status preference upon admission, but this process is loosely based and does not weigh a patient's values appropriately. By integrating ethicists into this conversation, patient values receive specific attention at the outset of an admission [171].

Similarly, clinical ethics can have a significant impact on compliance and risk management departments. These departments typically work to protect the hospital or health care system from legal ramifications by ensuring the hospital is engaging in proper legal regulations. Additionally, these departments intervene when there are specific instances where the hospital is at risk for legal repercussions or threats. The legitimacy of a department depends on its overall utility, and clinical ethics departments are no different [172]. To justify clinical ethicists' professionalization-with the components mentioned in this analysis, they need to demonstrate that their expertise directly impacts hospital departments that regulate issues like cost, risk, and overall patient care. While patient care seems like an obvious benefit of clinical ethics, the measurable impact of clinical ethics in departments like palliative medicine is of the utmost importance if professionalization occurs [173].

A clinical ethicist equipped with the critical virtue components mentioned in this analysis may provide significant insight into seemingly unreconcilable disputes. Implementing an ethicist in crucial conversations at the end of life can unearthing values and morally acceptable actions.

This level of involvement indicates yet another positive impact professional ethics can have on a health care system. However, viewing clinical ethics as a professional discipline and consultation services still faces various challenges [174]. Professionalizing clinical ethics may be accomplished through various means. While this analysis has offered ways in which curriculum points can aid the overall professionalization of clinical ethicists throughout their training, the means of viewing clinical ethics as a normative consultation service is still a glaring issue. This analysis argues that ethics should be viewed and utilized like any other medical consultation service, *i.e.,* gastroenterology, cardiology, etc. However, this tactic begs the question of whether clinical ethicists should bill for their services [175]. In American health care, individual consultants bill for services provided, including consultations. Typically, requestors place consultation orders if an attending physician determines that another professional's expertise is needed. The individual consultant will provide expertise and subsequently bill the patient for their services. In this respect, ethics consultants are no different from other medical professionals. However, differences in practice may allude to the advantages and disadvantages of billing for clinical ethics consultations [176].

Clinical ethics consultation can only become a professional vocation if medical culture views clinical ethicists as a legitimate consultation service in hospital settings. Billing for medical services legitimizes a professional discipline for various reasons. First, billing for professional services demonstrates the importance of a specific practice. Billing for services rendered indicates the prowess and knowledge provided by a professional [177]. Additionally, the scope and practice of a discipline that bills for its services indicate the level of importance the practice provides. Second, billing for clinical ethics consultation aids an overall institution's resources by obtaining funds from insurance companies and patients. Hospital funding and comprehensive resources for a hospital system are greatly improved if an ethics consultation service bills its services. Still, many individuals believe ethics consultants should not bill for their service. Billing for services is a contentious point due to the current position clinical ethics holds in hospital settings, the efforts to shift positions amongst the medical community, and the overall justification for an ethics department's existence.

While billing for clinical ethics has its advantages, specifically, the potential to promote clinical ethics consultants as professionals in health care, there are various drawbacks to billing patients and insurance companies for clinical ethics consultations. First, implementing billable practice for clinical ethics consultation can yield negative connotations for ethics consultants. Recently, various consultants in health care have abused billing systems. Ordering unnecessary tests, consultations, and medication orders for patients have fueled abuse throughout American health care. While medical professionals who call upon ethics consultants do not reap the benefits of billing for these services, receiving ethics consultations from patients and their families has the potential to plummet if these individuals become indebted for calling upon this service [178]. Second, ethics consultation services accompany patient rights. Within hospital policies, patient rights chapters typically indicate the rights that belong to patients upon admission to the hospital. Access to chaplain services, translation services, and ethics consultation services should accompany patient rights.

Billing for ethics consultation services inevitably yields more revenue for a hospital and ethics consultants. Still, the monetary facet of ethics assistance defeats the scope and purpose of the consult in the first place. While an attending physician may order services from other service lines and subsequently bill patients for those services, ethics consults may come from a physician, team member, family member, or patient. Knowing a patient requires diagnostic imaging or surgical intervention is a motivating action that may justify monetary duty. However, an ethicist's expertise should not abscond from a clinical situation due to financial constraints. Many ethical instances that arise in hospital medicine may surround a financial issue, resource availability issue, and other compounding factors that would dissuade a family or patient from seeking professional ethics advice. The overall revenue of a hospital can provide additional services and more full-time employees for hospital departments [179]. Still, the current climate of vocational ethics cannot afford to dissuade those in need of ethics services due to financial matters.

Billing for ethics consultation services infringes upon a patient's inalienable rights since billing for services involves placing a monetary restriction on individuals who cannot afford certain services [180]. Third, it is unjust to claim that ethics consultation services are available to all individuals in a health care organization if only one party is responsible for paying for the service. For example, a physician may request an ethics consultation for anobtunded patient without an available surrogate decision-maker. However, even if the patient's ethical discrepancy does not resolve with the ethicist's involvement, the patient is left with financial responsibility, including increased medical premiums or out-of-pocket payments if the patient does not have health insurance coverage [181].Although an option for legitimizing the profession, billing for ethics consultation services places a more significant burden on patients and health care than a benefit. The primary issue concerns patients' right to a service otherwise accessible by hospital employees and non-hospital employees alike.

While not billing for ethics consultation services is an issue in professionalizing the field, it is still possible to legitimize clinical ethics consultation without billing its services [182]. By implementing the key curriculum points presented in this analysis, the following analytic moral reasoning skills articulated in this analysis aid the legitimization of the profession without placing a financial burden upon involved stakeholders. Demonstrating the effectiveness of clinical ethics consultants with the curriculum points in this analysis legitimizes consultants' practice without requiring economic justification. Among other various issues, billing for ethics consultations still poses many difficulties, specifically in American health care. To address the difficulties accompanying the implementation of new curriculum methods for clinical ethicists, examining additional contemporary efforts in hospital-based bioethics programs may alleviate some of this discussion's practical concerns [183].

This discussion emphasizes key curriculum components that should be added to American bioethics programs to professionalize the discipline. It is highly beneficial to examine existing efforts of implementing hospital-based bioethics programs. The most formidable effort made derives from Renzo Pegoraro, Giovanni Putoto, and Emma Wray's efforts to establish hospital-based bioethics programs for European countries (EHBP). Examining Pegoraro, Putoto, and Wray's program intends to assess the provision of bioethics education in European countries and subsequently implement a model of bioethics courses in these participating countries. By examining their techniques, a greater understanding of how the proposed curriculum components in this analysis can integrate into American bioethics [184].

## A European Hospital-Based Bioethics Program

The European hospital-based bioethics program model established in Pegoraro, Putoto, and Wray's text, *Hospital Based Bioethics: A European Perspective*— henceforth referred to as Pegoraro's project—was first inspired by the advances in science and technology and the changes these advances have had on health care [185]. The increased benefits of technology and science in health care have arguably presented many ethical concerns. While hospitals serve as a venue that promotes collaboration of multidisciplinary expertise, a troubling dichotomy exists when an amalgam of professionals gather in this type of venue. The collaboration of expertise yields the potential for beneficial advancements in medicine and medical practice. However, there also exists the potential for clinical dilemmas [186].

Pegorarotackles this troubling dichotomy by establishing The European Hospital Bioethics Program (EHBP). Pegoraro developed his program with a bioethics team from ten European countries: France, Germany, Italy, Lithuania, the Netherlands, Poland, Portugal, the Slovak Republic, Slovenia, and the United Kingdom. By first identifying the objectives of these countries and the EHBP, a greater understanding of Pegoraro's project and the beneficial components for establishing a professional training program for American bioethicists. Additionally, examining the objectives and qualities that comprise the EHBP can grant more extraordinary insight into the applicability and effectiveness of the proposed curriculum components presented in this analysis [187]. Pegoraro presents five primary objectives. These objectives intend to demonstrate the commonality of bioethical goals between the ten countries that comprise the EHBP and solidify the ideal curriculum these countries would like to see in a hospital-based bioethics program. First, the EHBP seeks to assess the accessibility of bioethics education in hospitals. This objective intends to develop baseline data and aids in assessing the overall need for Pegoraro's project. Specifically, contrary to the ASBH's efforts, Pegoraro's approach intends to assess the broad range of western and central-eastern European countries' needs for a hospital-based bioethics program [188]. Astutely, Pegoraro's program attempts to assess the needs of these countries by referring to the contents and methods of existing programs. Second, Pegoraro's project uses the baseline data assessment from objective one to develop a bioethics course model tested in participating countries. Third, Pegoraro's project turns its focus to hospitals as the main venues for bioethics education. This method is highly relevant to implementing clinical bioethics programs since hospitals serve as ideal venues for disseminating clinical bioethics and care for many populations [189].

The first three objectives of Pegoraro's project demonstrate the assessment tactics and justification for hospital-based bioethics programs. Although published in 2007, this tactic is not an entirely novel endeavor. Since the move toward bioethical practice in medicine in the 1960s, medical practitioners, philosophers, and theologians alike developed various strategies and methodologies to disseminate ethics education to providers, patients, and other involved stakeholders. However, isolating expertise in a singularity has been a challenging endeavor. This analysis asserts that Pegoraro, the ASBH, and the EHBP's efforts share their objectives and scope of practice. The fourth objective of Pegoraro's project exemplifies this point by seeking to establish a common framework on clinical bioethics designed explicitly for hospitals in European Union countries and countries of Central-Eastern Europe [190]. While the ASBH's attempts at a bioethics program for clinical ethicists focus on American bioethics, the overall theme of establishing a methodology and education framework for a concentrated group of hospitals remains consistent. Naturally, the differences between European and American medicine vary greatly. Still, the ethical interventions share between these two camps are consistent and, for the most part, agreeable with one another. While bioethics methods inevitably require malleability between these camps due to the varying cultural practice of medicine, both camps understand that bioethics requires flexibility in its practice [191].

Finally, the fifth objective of Pegoraro's project aims to create a network of hospitals that are focused and dedicated to establishing an interactive bioethics forum. This objective is unique and lacking in American bioethics efforts. Since sharing ideas promotes the convergence of information, it is only practical to share findings with neighboring hospitals in like-minded health care systems. While the United States attempts to accomplish this objective by hosting bioethics conferences, forums, and collaborative conventions, there is no unifying method of bioethics education to share with neighboring hospital systems. In this respect, Pegoraro and the EHBP's networking bioethics methods are an excellent catalyst for establishing a unified methodology for teaching hospital-based bioethics. While there inherently lie differences in culture that do not allow a 1:1 conversion or sharing of bioethical teaching methods, there are significant benefits the ASBH and EHBP can gain from one another [192].

Unlike the ASBH and EHBP, the analyses presented in this work do not attempt to develop a hospital-based curriculum. Instead, this analysis attempts to introduce key components for an ethics consultation curriculum. While these curriculum points may manifest in academic circles, they intend to apply to various educational venues, including hospital-based programs. To develop a greater understanding of the EHBP's task and the applicability of the key curriculum points mentioned in this analysis, a deeper examination into Pegoraro's project and his shift from deontological ethics help explain the overall function and benefit virtue ethics and the development of analytic moral reasoning skills [193].

The objectives mentioned in Pegoraro and the EHBP's model are practical for European bioethics education in hospitals, but inevitable cultural changes must occur in American hospital systems if these methods should come to fruition. American bioethics has deep-seated roots in traditional deontological methods of moral evaluation. In his analysis, Pegoraro identifies this issue in European bioethics, too. The traditional deontological structure Pegoraro describes is an integral part of the inception of modern vocational bioethics. However, Pegoraro notes that adjustments and amendments to deontological structures in bioethics are necessary to form a formidable and unified method of hospital-based bioethics education [194].

Focusing upon innate rules and principles that govern and guide clinical decision-making has a relevant function. However, the same rigidity behind these rules can cause issues when adapting to the changing tides of health care. Pegoraro emphasizes this point in his first chapter by discussing the human rights movement's impact on global health care. While deontology initially held its roots in moral justification, Pegoraro argues that the deontological structure instituted by Immanuel Kant has transformed into medically relevant deontology [195]. Events like establishing the European Union Code on Medical Ethics have presented expanded principles upon which practitioners and ethicists alike ought to become familiar due to the changing tide of medical expertise and technological advances. While this is not a shift from deontology, establishing new principles aids in professionalizing bioethics as a normative discipline. Many scholars argue that the adaptation of principles that Pegoraro emphasizes is the very function deontology serves. However, shifts in deontological structures and principles have helped shape Pegoraro's project and thus aided in forming the EHBP's hospital-based bioethics program [196].

Additional points that Pegoraro mentions that indicate the shift in deontological structures in bioethics education entail the establishment of the American Medical Association's Code of Ethics, the introductions to informational technology, the Bioethics Committee of the Council of Europe (CDBI), and the European Group on Ethics (EGE). All the examples presented by Pegoraro's analysis demonstrate the need to adjust principles founded upon deontological foundations to adapt to changing tides in bioethical reasoning and thus solidify a progressive education program for hospital-based ethics [197]. Specifically, Pegoraro notes that the changing tide primarily drives these adjustments in technological advancements in health care. Pegoraro notes that this shift transforms deontology into medical deontology. Medical deontology services as a specific guiding principalistic method for abiding by rules. However, this shift also occurred due to the drastic change influenced by the human rights movement. Pegoraro notes that the European Convention of Human Rights (ECHR) tremendously influenced the dichotomy between legality and ethically fortuitous approaches [198]. The shift from legality to a deontological structure that accommodated impending principles led to the establishment of European International Legal Instruments in the Field of Biomedicine. These instruments intend to establish a connection between human rights and health care and combine them with the care context that human rights stand for. Pegoraro presents fundamental rights that are established based on legal provisions of bioethics:

1. Human beings have a right to be treated with respect and dignity
2. Human beings have the right to the highest attainable standard of physical and mental health
3. Human beings have the right to consent or to refuse medical interventions, including those related to research
4. Human beings have the right to protect against arbitrary interference with privacy or with family
5. Human beings have the right to enjoy the benefits of scientific progress and its application
6. Human beings have a right to protection for their rights, especially for vulnerable persons

The six rights Pegoraro presents are not inherently different from the rights outlined by American bioethics efforts. However, in American bioethics, these rights are often presented individually per a hospital's patient's rights guidelines [199]. Pegoraro's point is that these rights ought to be inherently entailed within the preview of bioethical practice. The rights mentioned above are not utterly alien to American bioethics. However, due to the nature and scope of the cultural differences between European and American bioethics, the above-mention rights, although ideal, do not abide by the cultural norms in American biomedical ethics. With the rights mentioned above in mind, Pegoraro notes that the process of understanding the role and place of bioethics in health care requires examinations at other normative approaches that have shaped the relationship between health care providers, society, and ethicists [200].

The two final points that Pegoraro notes in his assessment regard tradition and bioethics' power on other disciplines. First, the professional codes of medicine and ethics provide a historical background for developing a modern clinical ethics program for budding clinical ethicists. The legal, professional, and moral aspects of this facet of development must become inherently entailed in developing a formidable hospital-based bioethics program. However, bioethics as a traditional discipline, according to Pegoraro, trumps the secondary and tertiary aspects of ethics mention in his analysis. Second, Pegoraro notes that philosophy and other humanities have a tremendous role in establishing a formidable bioethical education program in hospital-based ethics. In this respect, Pegorara's analysis bolsters the need and effectiveness of the key components for an ethics consultation curriculum mentioned in this analysis [201].

It is essential to note that Pegoraro's attempts at creating a viable hospital-based bioethics program are successful in the context he proposes. While every facet of his approach may not become strictly applicable to American-based bioethics programs, the key curriculum points mentioned in this analysis have the potential to be applied to both European and American bioethics curriculums, despite the original derivation of curriculum points from an American bioethics perspective. The key components of virtue listed in this analysis that ultimately develop analytic moral reasoning skills for clinical ethicists aid Pegoraro's program by implementing a method that shifts from the deontological framework Pegoraro criticizes. Although categorized under applied program issues in this analysis, Pegoraro and the EHBP's efforts aid this discussion's efforts in accepting new curriculum points for hospital-based bioethics programs. By implementing Pegoraro's methods, the inherent morality of clinicians may be examined further and aid this analysis' overall goal of implementing uniform curriculum points for budding clinical ethicists [202].

Professional ethics within organizations is of the utmost importance in contemporary health care. Organizational ethics serves as a venue that formally introduces, collaborates, and professionalizes health care [203]. While the organizational aspect of professional health care is vital for the overall organization and structure of a health care institution, the ethical components associated with a health care organization further bolster the mission, identity, and core values that aid an organization's flourishing [204]. This analysis attempts to demonstrate the effectiveness of virtuous practices in health care organizations to examine the possibility of establishing an ethics consultation training curriculum that supports the curriculum points mentioned in this analysis.

By implementing virtue ethics throughout a health care organization, analytic moral reasoning skills are developed and honed throughout the system. Fostering these skills has an array of positive outcomes for the entire organization [205]. By training individuals to identify instances of excess and deficiencies of virtue, individuals inevitably ascertain instances of virtuous practice by understanding its antithesis. Identifying virtue and its deficiencies or excesses cultivates an organizational atmosphere that promotes moral agency throughout its system [206]. This process completes by introducing virtue-identification techniques as mandatory competencies throughout a health care organization. Synonymous with the ASBH's core competencies for clinical ethics consultants, establishing virtue identification techniques as mandatory competencies promotes an ethical atmosphere and recognizes the mission and identity of an institution as the pinnacle of moral professionalism [207]. The subsequent moral agency derived from analytic moral reasoning and virtue identification facilitates conflict resolution techniques by bolstering moral cooperation. Furthermore, ethical leadership becomes a far more critical role due to the hierarchical nature of health care organizations and the influence upper management possesses [208].

Ideally, clinical ethicists are the most qualified individuals for facilitating virtuous behaviors in a health care system. Formal education in a residency program expedites the integration of virtue throughout an institution and subsequently develops an ethical culture throughout a hospital system. Integrating virtue ethics into organizational curriculums justifies the moral nature involved in clinical ethics and demonstrates the need to formalize ethics education in healthcare organizations and implement virtue ethics as a standard competency [209].

While a robust philosophical background is beneficial for clinical ethics consultants, it is impossible to expect a multidisciplinary field like bioethics to accommodate individuals who possess this prerequisite. Alternatively, a health care organization's educational emphasis on virtue-identification techniques yields a set of analytic moral reasoning skills that serve as a profound substitute for an extensive background in moral theory. These analytic moral reasoning skills are fundamental and infinitely valuable due to their ability to bolster and adopt practical competencies, promote ethical character, and further establish an organization's duties toward stewardship and integrity [210].

Residency programs indicate a formidable venue for hosting the critical components mentioned in this analysis. Still, various issues accompany this task. First, the cultural attitude in medicine toward clinical ethicists continues to thwart professional efforts in establishing a comprehensive ethics residency program. The analysis indicates that this issue, among others, can resolve if the sole-consultant model adopts in hospital medicine. The sole-consultant model expedites a cultural shift in medicine by palliating the fears of health care professionals and educating professionals about the role and scope of clinical ethics.

Additionally, issues that inhibit ethical presences in hospital settings include the lack of information other departments possess concerning ethics consultation. While clinical ethics consultants typically focus on bedside ethics, there are various venues where ethics expertise can help an entire hospital system. Compliance, risk management, and quality improvement departments can benefit from clinical ethics expertise since patient care is at the center of their practice. Identifying the gaps in knowledge surrounding clinical ethics knowledge and implementing this information to a more extensive hospital system aids the effort in establishing clinical ethics residency programs in health care systems by assisting a cultural paradigm shift.

Finally, this analysis's final points introduce existing efforts in implementing hospital-based ethics programs in Europe. By examining Renzo Pegoraro's attempts to assess and establish a hospital-based ethics program in Europe, the ASBH can learn tremendously in its current efforts. Methods of extracting and assessing data regarding the demands of clinical ethics are an excellent way of beginning this process. Additionally, assessing the philosophical paradigms a program of this magnitude exemplifies aids the American effort by extracting effective methods of introducing ethics to an otherwise resistant system. However, the issues accompanying an American model of hospital-based bioethics programs from Pegoraro's program lie in cultural differences in medicine and health care. While Pegoraro and the EHBP's efforts have proven effective in the ten participating European countries, establishing a clinical ethics residency program that caters to the current climate of healthcare in the United States becomes an onerous process that may never receive attention.

While the purpose of this discussion is not to establish a novel curriculum, nor is it to develop a school for ethics, it intends to introduce a small yet powerful educational direction that ethicists must direct their attention in their training. Identifying ethics consultation programs aids this process by examining if introducing these curriculum components is even tenable. However, this discussion argues that, despite the cultural issues surrounding clinical ethics, the critical curriculum components mentioned in this analysis inevitably bolster training programs for ethicists positively and pragmatically.

# References

1. Nielsen, Richard P., and Felipe G. Massa. "Reintegrating Ethics and Institutional Theories." Journal of Business Ethics 115, no. 1 (2012): 588-596.
2. Beauchamp, Tom L., and James F. Childress, *Principles of biomedical ethics.* 6th ed. (New York: Oxford University Press, 2016), 240-280.
3. Treviño, L. K., Weaver, G. R., & Brown, M. E., It's Lovely at the Top: Hierarchical Levels, Identities, and Perceptions of Organizational Ethics. Business Ethics Quarterly 18 no. 02, (2008): 247-249.
4. Perry, Michael J. *The Idea of Human Rights: Four Inquiries.* New York, NY: Oxford University Press, 1998. 3-5.
5. Gallagher, John A., and Jerry Goodstein. "Fulfilling Institutional Responsibilities in Health Care: Organizational Ethics and the Role of Mission Discernment." Business Ethics Quarterly 12, no. 4 (2002). 433-435.
6. Magill, Gerard, and Lawrence Prybil. "Stewardship and Integrity in Health Care: A Role for Organizational Ethics." Journal of Business Ethics 50, no. 3 (2004). 227-228.
7. Magill, Gerard, and Lawrence Prybil. "Stewardship and Integrity in Health Care: A Role for Organizational Ethics." Journal of Business Ethics 50, no. 3 (2004). 228.
8. Magill, Gerard, and Lawrence Prybil. "Stewardship and Integrity in Health Care: A Role for Organizational Ethics." Journal of Business Ethics 50, no. 3 (2004). 230-233.
9. Gallagher, John A., and Jerry Goodstein. "Fulfilling Institutional Responsibilities in Health Care: Organizational Ethics and the Role of Mission Discernment." Business Ethics Quarterly 12, no. 4 (2002). 434.
10. Gallagher, John A., and Jerry Goodstein. "Fulfilling Institutional Responsibilities in Health Care: Organizational Ethics and the Role of Mission Discernment." Business Ethics Quarterly 12, no. 4 (2002). 434-440.
11. Pellegrino, Edmund D., and David C. Thomasma. A philosophical basis of medical practice: Toward a Philosophy and Ethic of the Healing Professions. New York: Oxford University Press, 1981. 39-47
12. Pellegrino, Edmund D., and David C. Thomasma. A philosophical basis of medical practice: Toward a Philosophy and Ethic of the Healing Professions. New York: Oxford University Press, 1981. 244-246.
13. Nielsen, Richard P. "Can ethical character be stimulated and enabled? An action-learning approach to teaching and learning organization ethics." *Research in Ethical Issues in Organizations The Next Phase of Business Ethics: Integrating Psychology and Ethics.* 595-600.
14. Calabrese, Raymond L., and Angela Calabrese Barton. "The Practice of Integrity Within the University." *The Journal of Educational Thought*, no. 3 (December 2000): 270-275.
15. Calabrese, Raymond L., and Angela Calabrese Barton. "The Practice of Integrity Within the University." *The Journal of Educational Thought*, no. 3 (December 2000): 266-268.
16. Arendt, Hannah, and Margaret Canovan. The Human Condition. Chicago: University of Chicago Press, 2012. 50-58.
17. Calabrese, Raymond L., and Angela Calabrese Barton. "The Practice of Integrity Within the University." *The Journal of Educational Thought*, no. 3 (December 2000): 268-270.
18. Calabrese, Raymond L., and Angela Calabrese Barton. "The Practice of Integrity Within the University." *The Journal of Educational Thought*, no. 3 (December 2000): 268-269.
19. Magill, Gerard, and Lawrence Prybil. "Stewardship and Integrity in Health Care: A Role for Organizational Ethics." Journal of Business Ethics 50, no. 3 (2004). 225-227.
20. Beauchamp, Tom L., and James F. Childress, *Principles of biomedical ethics.* 6th ed. (New York: Oxford University Press, 2016), 240-280.
21. Boyle, Philip. Organizational ethics in health care: principles, cases, and practical solutions. San Francisco: Jossey-Bass, 2001. 247-248.

22. Gallagher, John A., and Jerry Goodstein. "Fulfilling Institutional Responsibilities in Health Care: Organizational Ethics and the Role of Mission Discernment." Business Ethics Quarterly 12, no. 4 (2002). 435-440.

23. "Mission, Core Values and Vision." Trinity Health - Livonia, Michigan (MI) Hospitals. http://www.trinity-health.org/mission-values.

24. Gallagher, John A., and Jerry Goodstein. "Fulfilling Institutional Responsibilities in Health Care: Organizational Ethics and the Role of Mission Discernment." Business Ethics Quarterly 12, no. 4 (2002). 445-450.

25. Sekerka, Leslie E., Richard P. Bagozzi, and Richard Charnigo. "Facing Ethical Challenges in the Workplace: Conceptualizing and Measuring Professional Moral Courage." Journal of Business Ethics 89, no. 4 (2009). 570-578.

26. Calabrese, Raymond L., and Angela Calabrese Barton. "The Practice of Integrity Within the University." *The Journal of Educational Thought*, no. 3 (December 2000): 275-280.

27. O'Kelly, C. "Taking Tough Choices Seriously: Public Administration and Individual Moral Agency." Journal of Public Administration Research and Theory 16, no. 3 (2005). 396-398.

28. Magill, Gerard, and Lawrence Prybil. "Stewardship and Integrity in Health Care: A Role for Organizational Ethics." Journal of Business Ethics 50, no. 3 (2004). 232-235.

29. Calabrese, Raymond L., and Angela Calabrese Barton. "The Practice of Integrity Within the University." *The Journal of Educational Thought*, no. 3 (December 2000): 268-270.

30. Magill, Gerard, and Lawrence Prybil. "Stewardship and Integrity in Health Care: A Role for Organizational Ethics." Journal of Business Ethics 50, no. 3 (2004). 229-232.

31. Magill, Gerard, and Lawrence Prybil. "Stewardship and Integrity in Health Care: A Role for Organizational Ethics." Journal of Business Ethics 50, no. 3 (2004). 232-235.

32. Calabrese, Raymond L., and Angela Calabrese Barton. "The Practice of Integrity Within the University." *The Journal of Educational Thought*, no. 3 (December 2000): 266-267.

33. Pellegrino, Edmund D., and David C. Thomasma. A philosophical basis of medical practice: Toward a Philosophy and Ethic of the Healing Professions. New York: Oxford University Press, 1981. 244.

34. Block, Fred. "A Corporation With A Conscience?" New Labor Forum 15, no. 2 (2006): 75-83.

35. Verbos, Amy Klemm, Joseph A. Gerard, Paul R. Forshey, Charles S. Harding, and Janice S. Miller. "The Positive Ethical Organization: Enacting a Living Code of Ethics and Ethical Organizational Identity." *Journal of Business Ethics* 76, no. 1 (2007). 30-35.

36. Perry, Michael J. *The Idea of Human Rights: Four Inquiries*. New York, NY: Oxford University Press, 1998. 7-15.

37. American Society for Bioethics and Humanities. 2011. Core Competencies for Health Care Ethics Consultation, 2nd edition. Glenview, IL.

38. Treviño, L. K., Weaver, G. R., & Brown, M. E., It's Lovely at the Top: Hierarchical Levels, Identities, and Perceptions of Organizational Ethics. Business Ethics Quarterly 18 no. 02, (2008): 233-237.

39. Treviño, L. K., Weaver, G. R., & Brown, M. E., It's Lovely at the Top: Hierarchical Levels, Identities, and Perceptions of Organizational Ethics. Business Ethics Quarterly 18 no. 02, (2008): 238-243.

40. Treviño, L. K., Weaver, G. R., & Brown, M. E., It's Lovely at the Top: Hierarchical Levels, Identities, and Perceptions of Organizational Ethics. Business Ethics Quarterly 18 no. 02, (2008): 247-249.

41. Boyle, Philip. Organizational ethics in health care: principles, cases, and practical solutions. San Francisco: Jossey-Bass, 2001. 326-328.

42. Boyle, Philip. Organizational ethics in health care: principles, cases, and practical solutions. San Francisco: Jossey-Bass, 2001. 330.

43. Sekerka, Leslie E., Richard P. Bagozzi, and Richard Charnigo. "Facing Ethical Challenges in the Workplace: Conceptualizing and Measuring Professional Moral Courage." Journal of Business Ethics 89, no. 4 (2009). 567-570.

44. Pellegrino, Edmund D., and David C. Thomasma. A philosophical basis of medical practice: Toward a Philosophy and Ethic of the Healing Professions. New York: Oxford University Press, 1981. 250.
45. Valentine, Sean, and Tim Barnett. "Perceived Organizational Ethics and the Ethical Decisions of Sales and Marketing Personnel." Journal of Personal Selling and Sales Management 27, no. 4 (2007). 380.
46. McLean, Sheila, and Gerard Magill. First do no harm: law, ethics and healthcare. Aldershot, England: Ashgate, 2008. 106-108.
47. O'Kelly, C. "Taking Tough Choices Seriously: Public Administration and Individual Moral Agency." Journal of Public Administration Research and Theory 16, no. 3 (2005). 396-398.
48. Sekerka, Leslie E., Richard P. Bagozzi, and Richard Charnigo. "Facing Ethical Challenges in the Workplace: Conceptualizing and Measuring Professional Moral Courage." Journal of Business Ethics 89, no. 4 (2009). 565-566.
49. American Society for Bioethics and Humanities. 2011. Core Competencies for Health Care Ethics Consultation, 2nd edition. Glenview, IL.
50. Sekerka, Leslie E., Richard P. Bagozzi, and Richard Charnigo. "Facing Ethical Challenges in the Workplace: Conceptualizing and Measuring Professional Moral Courage." Journal of Business Ethics 89, no. 4 (2009). 567-568.
51. Sekerka, Leslie E., Richard P. Bagozzi, and Richard Charnigo. "Facing Ethical Challenges in the Workplace: Conceptualizing and Measuring Professional Moral Courage." Journal of Business Ethics 89, no. 4 (2009). 568-572.
52. Nielsen, Richard P. "Can ethical character be stimulated and enabled? An action-learning approach to teaching and learning organization ethics." *Research in Ethical Issues in Organizations The Next Phase of Business Ethics: Integrating Psychology and Ethics.* 587-593.
53. Nielsen, Richard P. "Can ethical character be stimulated and enabled? An action-learning approach to teaching and learning organization ethics." *Research in Ethical Issues in Organizations The Next Phase of Business Ethics: Integrating Psychology and Ethics.* 595-600.
54. Dempsey, James. "Corporations and Non-Agential Moral Responsibility." Journal of Applied Philosophy 30, no. 4 (2013).
55. Mcdonald, Ross A., and Bart Victor. "Towards the Integration of Individual and Moral Agencies." Business and Professional Ethics Journal 7, no. 3 (1988). 103-107.
56. Pellegrino, Edmund D., and David C. Thomasma. A philosophical basis of medical practice: Toward a Philosophy and Ethic of the Healing Professions. New York: Oxford University Press, 1981. 244.
57. Lynch, Thomas D., and Cynthia E. Lynch. "Virtue Ethics: A Policy Recommendation." *Public Administration Quarterly*, Winter 2002, 25, no. 4 (2002): 462-471.
58. Nielsen, Richard P. "Can ethical character be stimulated and enabled? An action-learning approach to teaching and learning organization ethics." *Research in Ethical Issues in Organizations The Next Phase of Business Ethics: Integrating Psychology and Ethics.* 581-587.
59. Oosterhout, J. Van, Ben Wempe, and Theo Van Willigenburg. "Rethinking Organizational Ethics: A Plea for Pluralism." Journal of Business Ethics 55, no. 4 (2004): 387-389.
60. Mcdonald, Ross A., and Bart Victor. "Towards the Integration of Individual and Moral Agencies." Business and Professional Ethics Journal 7, no. 3 (1988). 103-104.
61. Treviño, L. K., Weaver, G. R., & Brown, M. E., It's Lovely at the Top: Hierarchical Levels, Identities, and Perceptions of Organizational Ethics. Business Ethics Quarterly 18 no. 02, (2008): 240-242.
62. Lynch, Thomas D., and Cynthia E. Lynch. "Virtue Ethics: A Policy Recommendation." *Public Administration Quarterly*, Winter 2002, 25, no. 4 (2002): 471-472.
63. Murphy, Patrick E. "Character and Virtue Ethics in International Marketing: An Agenda for Managers, Researchers and Educators." *Journal of Business Ethics* 18, no. 1 (January 1999). 108-114.

64. Sims, Ronald R. "The institutionalization of organizational ethics." *Journal of Business Ethics* 10, no. 7 (1991). 498-505.
65. McLean, Sheila, and Gerard Magill. First do no harm: law, ethics and healthcare. Aldershot, England: Ashgate, 2008. 106-108.
66. McLean, Sheila, and Gerard Magill. First do no harm: law, ethics and healthcare. Aldershot, England: Ashgate, 2008. 108-111.
67. Mcdonald, Ross A., and Bart Victor. "Towards the Integration of Individual and Moral Agencies." Business and Professional Ethics Journal 7, no. 3 (1988). 103-107.
68. Boyle, Philip. Organizational ethics in health care: principles, cases, and practical solutions. San Francisco: Jossey-Bass, 2001. 292-295.
69. O'Kelly, C. "Taking Tough Choices Seriously: Public Administration and Individual Moral Agency." Journal of Public Administration Research and Theory 16, no. 3 (2005). 393-396.
70. O'Kelly, C. "Taking Tough Choices Seriously: Public Administration and Individual Moral Agency." Journal of Public Administration Research and Theory 16, no. 3 (2005). 396-398.
71. Gallagher, John A., and Jerry Goodstein. "Fulfilling Institutional Responsibilities in Health Care: Organizational Ethics and the Role of Mission Discernment." Business Ethics Quarterly 12, no. 4 (2002). 444-446.
72. O'Kelly, C. "Taking Tough Choices Seriously: Public Administration and Individual Moral Agency." Journal of Public Administration Research and Theory 16, no. 3 (2005). 399-401.
73. Valentine, Sean, and Tim Barnett. "Perceived Organizational Ethics and the Ethical Decisions of Sales and Marketing Personnel." Journal of Personal Selling and Sales Management 27, no. 4 (2007). 373-375.
74. Lynch, Thomas D., and Cynthia E. Lynch. "Virtue Ethics: A Policy Recommendation." *Public Administration Quarterly*, Winter 2002, 25, no. 4 (2002): 464-472.
75. Lynch, Thomas D., and Cynthia E. Lynch. "Virtue Ethics: A Policy Recommendation." *Public Administration Quarterly*, Winter 2002, 25, no. 4 (2002): 471-472.
76. Lynch, Thomas D., and Cynthia E. Lynch. "Virtue Ethics: A Policy Recommendation." *Public Administration Quarterly*, Winter 2002, 25, no. 4 (2002): 473-474.
77. Lynch, Thomas D., and Cynthia E. Lynch. "Virtue Ethics: A Policy Recommendation." *Public Administration Quarterly*, Winter 2002, 25, no. 4 (2002): 475-481.
78. Hartman, Edwin M. "Moral Philosophy, Political Philosophy, and Organizational Ethics: A Response to Phillips and Margolis." *Business Ethics Quarterly* 11, no. 4 (2001): 679-683.
79. Weber, Todd Bernard. "Analyzing Wrongness as Sanction-Worthiness." *The Journal of Value Inquiry* 40, no. 1 (2007). 23-29.
80. Weber, Todd Bernard. "Analyzing Wrongness as Sanction-Worthiness." *The Journal of Value Inquiry* 40, no. 1 (2007). 29-30.
81. Cafardi, Nicholas P., and Gerard Magill. Voting and holiness: Catholic perspectives on political participation. New York: Paulist Press, 2012. 135-136.
82. Cafardi, Nicholas P., and Gerard Magill. Voting and holiness: Catholic perspectives on political participation. New York: Paulist Press, 2012. 136-137.
83. Cafardi, Nicholas P., and Gerard Magill. Voting and holiness: Catholic perspectives on political participation. New York: Paulist Press, 2012. 139-141.
84. Cafardi, Nicholas P., and Gerard Magill. Voting and holiness: Catholic perspectives on political participation. New York: Paulist Press, 2012. 140-142.
85. Moore, Geoff. "Corporate Character: Modern Virtue Ethics and the Virtuous Corporation." Business Ethics Quarterly 15, no. 04 (2005). 660-662.
86. Moore, Geoff. "Corporate Character: Modern Virtue Ethics and the Virtuous Corporation." Business Ethics Quarterly 15, no. 04 (2005). 663-669.
87. Magill, Gerard, and Lawrence Prybil. "Stewardship and Integrity in Health Care: A Role for Organizational Ethics." Journal of Business Ethics 50, no. 3 (2004). 231-235.
88. Gan, Shaoping, and Lin Zhang. "The Destiny of Modern Virtue Ethics." *Frontiers of Philosophy in China*, Sept, 5, no. 3 (2010). 432–435.
89. American Society for Bioethics and Humanities. 2011. Core Competencies for Health Care Ethics Consultation, 2nd edition. Glenview, IL.

90. Vertrees, Stephanie M., Andrew G. Shuman, and Joseph J. Fins. "Learning by Doing: Effectively Incorporating Ethics Education into Residency Training." Journal of General Internal Medicine 28, no. 4 (2012): 570-575.
91. Vertrees, Stephanie M., Andrew G. Shuman, and Joseph J. Fins. "Learning by Doing: Effectively Incorporating Ethics Education into Residency Training." Journal of General Internal Medicine 28, no. 4 (2012): 578-579.
92. Vertrees, Stephanie M., Andrew G. Shuman, and Joseph J. Fins. "Learning by Doing: Effectively Incorporating Ethics Education into Residency Training." Journal of General Internal Medicine 28, no. 4 (2012): 580-582.
93. Marder, William D., and Douglas E. Hough. "Medical Residency as Investment in Human Capital." The Journal of Human Resources 18, no. 1 (1983). 50-52.
94. A Pilot Evaluation of Portfolios for Quality Attestation of Clinical Ethics Consultants." National Center for Biotechnology Information. U.S. National Library of Medicine, n.d. Web. 11 Nov. 2016. 1-16.
95. Gan, Shaoping. "The destiny of modern virtue ethics." Frontiers of Philosophy in China 5, no. 3 (2010). 435-431.
96. Lynch, Thomas D., and Cynthia E. Lynch. "Virtue Ethics: A Policy Recommendation." *Public Administration Quarterly*, Winter 2002, 25, no. 4 (2002): 493-495.
97. Lynch, Thomas D., and Cynthia E. Lynch. "Virtue Ethics: A Policy Recommendation." *Public Administration Quarterly*, Winter 2002, 25, no. 4 (2002): 485.
98. Nielsen, Richard P. "Can ethical character be stimulated and enabled? An action-learning approach to teaching and learning organization ethics." *Research in Ethical Issues in Organizations The Next Phase of Business Ethics: Integrating Psychology and Ethics.* 581-583.
99. Sims, Ronald R. "The institutionalization of organizational ethics." *Journal of Business Ethics* 10, no. 7 (1991): 493-495.
100. Carter, Michele A., and Craig M. Klugman. "Cultural Engagement in Clinical Ethics: A Model for Ethics Consultation." Cambridge Quarterly of Healthcare Ethics 10, no. 1, (2001): 301-333.
101. Sims, Ronald R. "The institutionalization of organizational ethics." *Journal of Business Ethics* 10, no. 7 (1991): 502-503.
102. Sims, Ronald R. "The institutionalization of organizational ethics." *Journal of Business Ethics* 10, no. 7 (1991): 504.
103. Sims, Ronald R. "The institutionalization of organizational ethics." *Journal of Business Ethics* 10, no. 7 (1991): 504.
104. Sims, Ronald R. "The institutionalization of organizational ethics." *Journal of Business Ethics* 10, no. 7 (1991): 503-505.
105. Rhodes, Carl, Alison Pullen, and Stewart R. Clegg. "'If I Should Fall From Grace…': Stories of Change and Organizational Ethics." Journal of Business Ethics 91, no. 4 (2009). 535-537.
106. Pellegrino, Edmund D., and David C. Thomasma. *The virtues in medical practice.* (New York u.a.: Oxford Univ. Press, 1993), 187-189.
107. Pellegrino, Edmund D., and David C. Thomasma. *The virtues in medical practice.* (New York u.a.: Oxford Univ. Press, 1993), 184-191.
108. Verbos, Amy Klemm, Joseph A. Gerard, Paul R. Forshey, Charles S. Harding, and Janice S. Miller. "The Positive Ethical Organization: Enacting a Living Code of Ethics and Ethical Organizational Identity." *Journal of Business Ethics* 76, no. 1 (2007). 20-22.
109. Verbos, Amy Klemm, Joseph A. Gerard, Paul R. Forshey, Charles S. Harding, and Janice S. Miller. "The Positive Ethical Organization: Enacting a Living Code of Ethics and Ethical Organizational Identity." *Journal of Business Ethics* 76, no. 1 (2007). 22-25.
110. Arendt, Hannah, and Margaret Canovan. The Human Condition. Chicago: University of Chicago Press, 2012. 50-58.

111. Verbos, Amy Klemm, Joseph A. Gerard, Paul R. Forshey, Charles S. Harding, and Janice S. Miller. "The Positive Ethical Organization: Enacting a Living Code of Ethics and Ethical Organizational Identity." *Journal of Business Ethics* 76, no. 1 (2007). 22-28.

112. Zajac, G., and L. K. Comfort. ""The Spirit of Watchfulness": Public Ethics as Organizational Learning." Journal of Public Administration Research and Theory 7, no. 4 (1997). 545-550.

113. Nelson, William, Gili Lushkov, Andrew Pomerantz, and William B. Weeks. "Rural health care ethics: Is there a literature?." The American Journal of Bioethics 6, no. 2 (2006): 44-50.

114. Pence, Gregory E. "Classic Cases in Medical Ethics: Accounts of Cases that Have Shaped Medical Ethics, with Philosophical, Legal, and Historical Backgrounds." (2004): 24-46.

115. Cartledge, Gwendolyn, Linda C. Tillman, and Carolyn Talbert Johnson. "Professional ethics within the context of student discipline and diversity," *Teacher Education and Special Education* 24, no. 1 (2001): 25-37.

116. O'neill, Onora. *Autonomy and trust in bioethics*, (Cambridge University Press, 2002), 28-44.

117. Wachter, Robert M., and Lee Goldman. "The emerging role of "hospitalists" in the American health care system." (1996): 514-517.

118. Lindenauer, Peter K., Michael B. Rothberg, Penelope S. Pekow, Christopher Kenwood, Evan M. Benjamin, and Andrew D. Auerbach. "Outcomes of care by hospitalists, general internists, and family physicians." New England Journal of Medicine 357, no. 25 (2007): 2589-2600.

119. Kenny, Dianna T., "Determinants of patient satisfaction with the medical consultation," Psychology and Health 10, no. 5 (1995): 427-437.

120. DuVal, Gordon, Leah Sartorius, Brian Clarridge, Gary Gensler, and Marion Danis. "What triggers requests for ethics consultations?," Journal of Medical Ethics 27, no. 1 (2001): 24-29.

121. Cho, Mildred K., Sara L. Tobin, Henry T. Greely, Jennifer McCormick, Angie Boyce, and David Magnus. "Strangers at the benchside: Research ethics consultation." The American Journal of Bioethics 8, no. 3 (2008): 4-13.

122. Danis, Marion, Adrienne Farrar, Christine Grady, Carol Taylor, Patricia O'Donnell, Karen Soeken, and Connie Ulrich. "Does fear of retaliation deter requests for ethics consultation?," Medicine, Health Care and Philosophy 11, no. 1 (2008): 27-34.

123. Self, Donnie J., Joy D. Skeel, and Nancy S. Jecker. "A comparison of the moral reasoning of physicians and clinical medical ethicists." Academic Medicine 68, no. 11 (1993): 852-855.

124. Schneiderman, Lawrence J., Todd Gilmer, and Holly D. Teetzel. "Impact of ethics consultations in the intensive care setting: a randomized, controlled trial." *Critical care medicine* 28, no. 12 (2000): 3920-3924.

125. Rushton, Cynda, Stuart J. Youngner, and Joy Skeel. "Models for Ethics Consultation: Individual, Team, or Committee?," (Baltimore: Johns Hopkins University Press, 2003), 88-95.

126. DuVal, Gordon, Brian Clarridge, Gary Gensler, and Marion Danis. "A national survey of U.S. internists' experiences with ethical dilemmas and ethics consultation." *Journal of general internal medicine* 19, no. 3 (2004): 251-258.

127. Fischer, Josie. "Social responsibility and ethics: clarifying the concepts," Journal of Business ethics 52, no. 4 (2004): 381-390.

128. McGee, Glenn, ed. *Pragmatic bioethics*, (MIT press, 2003), 1-44.

129. Banerjee, Dipanjan, and Ware G Kuschner, "Principles and procedures of medical ethics case consultation." British Journal of Hospital Medicine 68, no. 3 (2007): 140-144.

130. Jonsen, Albert R., Mark Siegler, and William J. Winslade. *Clinical Ethics: A Practical Approach to Ethical Decisions in Clinical Medicine*. 8th ed. (New York: McGraw Hill, Medical Pub. Division, 2015), 33-45.

131. Mackler, Aaron L. *Introduction to Jewish and Catholic Bioethics: A Comparative Analysis*. (Washington, D.C.: Georgetown University Press, 2003), 12-17.

132. Strong, Carson. "Gert's Moral Theory and Its Application to Bioethics Cases." Kennedy Institute of Ethics Journal 16, no. 1 (2006): 39-58.

133. Vertrees, Stephanie M., Andrew G. Shuman, and Joseph J. Fins. "Learning by Doing: Effectively Incorporating Ethics Education into Residency Training." Journal of General Internal Medicine 28, no. 4 (2012): 578-82.

134. Zahedi, F., M. Sanjari, M. Aala, M. Peymani, K. Aramesh, A. Parsapour, SS Bagher Maddah, MA Cheraghi, GH Mirzabeigi, B. Larijani, and M. Vahid Dastgerdi. "The Code of Ethics for Nurses." Iranian Journal of Public Health 42, no. 1 (2013): 1-8.
135. Pellegrino, Edmund D., and David C. Thomasma. *A Philosophical Basis of Medical Practice: Toward a Philosophy and Ethic of the Healing Professions.* (New York: Oxford University Press, 1981), 12-16.
136. Walker, Margaret Urban. "Keeping Moral Space Open New Images of Ethics Consulting." The Hastings Center Report 23, no. 2 (1993): 33.
137. Dubler, N. N., M. P. Webber, D. M. Swiderski, et al., "Charting the Future: Credentialing, Privileging, Quality, and Evaluation in Clinical Ethics Consultation," Hastings Center Report, 39 no. 6 (2009): 23-33.
138. McKneally, Martin F., and Peter A. Singer. "Bioethics for Clinicians: 25. Teaching Bioethics in the Clinical Setting," Canadian Medical Association or Its Licensors, (2001): 1163-1167.
139. Bernstein, Mark, and Kerry Bowman. "Should a Medical/Surgical Specialist with Formal Training in Bioethics Provide Health Care Ethics Consultation in His/her Own Area of Specialty?" HEC Forum.
140. Asghari, Fariba, Aniseh Samadi, and Arash Rashidian. "Medical Ethics Course for Undergraduate Medical Students: A Needs Assessment Study." Journal of Medical Ethics and History of Medicine 6, no. 7 (2013): 1-7.
141. Pearlman, R. A., M. Foglia, E. Fox, J. Cohen, B. L. Chanko, and K. Berkowitz. 2016. Ethics Consultation Quality Assessment Tool: A Novel Method for Assessing the Quality of Ethics Case Consultations Based on Written Records. American Journal of Bioethics, 16(3): 3-14.
142. Aulisio, M. P., R. M. Arnold, and S. J. Youngner, "Health Care Ethics Consultation: Nature, Goals, and Competencies," Annals of Internal Medicine, 1 No.133 (2000): 59-69.
143. Orr, R. D., and W. Shelton. "A process and format for clinical ethics consultation." Journal of Clinical Ethics 20 no. 1, (2009): 79-89.
144. Carrese, Joseph A. and the Members of the American Society for Bioethics and Humanities Clinical Ethics Consultation Affairs Standing Committee. "HCEC Pearls and Pitfalls: Suggested Do's and Don'ts for Healthcare Ethics Consultants." *Journal of Clinical Ethics,* 23, No. 3, (2012): 234-240.
145. Repenshek, Mark. "Quality Attestation for Clinical Ethics Consultants: Perspectives from the Field." Catholic Health Association. Saint Louis, MO. 2014.
146. Tinker, Anthea, and Vera Coomber. "University research ethics committees: Their role, remit and conduct." Bulletin of medical ethics 203 (2004): 7-8.
147. Fox, E., S. Myers, and R. A. Pearlman. 2007, "Ethics Consultation in United States Hospitals: A National Survey," *American Journal of Bioethics* 7 no. 2, (2007): 10-15.
148. Fletcher, John C., and Diane E. Hoffmann, "Ethics committees: Time to experiment with standards," Annals of Internal Medicine 120, no. 4 (1994): 335-338.
149. Weir, Michael. "Ethics and Professional Misconduct. Alternative Medicine: A New Regulatory Model," Notre Dame Law Review (2005): 260.
150. Førde, Reidun, and I. H. Vandvik, "Clinical ethics, information, and communication: review of 31 cases from a clinical ethics committee," *Journal of medical ethics* 31, no. 2 (2005): 73-77.
151. Fuchs, Michael, and Nationaler Ethikrat, *National ethics councils: their backgrounds, functions and modes of operation compared,* (Berlin: Nationaler Ethikrat, 2005): 100; *See* La Puma, John M. D., and David L. Schiedermayer, "Ethics consultation: skills, roles, and training," *Annals of Internal Medicine* 114 (1991): 155-160.
152. Engelhardt, H. Tristram. "Core competencies for health care ethics consultants: In search of professional status in a post-modern world," *HEC forum*, 23, no. 3, (Springer Netherlands, 2011): 129.
153. Grady, Christine, Marion Danis, Karen L. Soeken, Patricia O'Donnell, Carol Taylor, Adrienne Farrar, and Connie M. Ulrich. "Does ethics education influence the moral action of practicing nurses and social workers?." The American Journal of Bioethics 8, no. 4 (2008): 4-11.

154. Fallona, Catherine. "Manner in teaching: a study in observing and interpreting teachers' moral virtues." Teaching and Teacher Education 16, no. 7 (2000): 681-695.

155. Jamal, Tazim B. "Virtue ethics and sustainable tourism pedagogy: Phronesis, principles and practice." *Journal of sustainable tourism* 12, no. 6 (2004): 530-545.

156. Melé, Domènec. "Ethical education in accounting: Integrating rules, values and virtues." Journal of Business Ethics 57, no. 1 (2005): 97-109.

157. Harrison, John. "Conflicts of duty and the virtues of Aristotle in public relations ethics: Continuing the conversation commenced by Monica Walle." PRism 2, no. 1 (2004): 1-7.

158. Aristotle, and W. D. Ross. *The Nichomachean ethics*. London: Oxford University Press, (1959): 3.6-9.

159. Darr, Kurt. "Virtue ethics: worth another look." *Hospital topics*84, no. 4 (2006): 29-31.

160. Nussbaum, Martha C. "Non-relative virtues: an Aristotelian approach." *Midwest studies in philosophy* 13, no. 1 (1988): 32-53.

161. Coulehan, Jack, and Peter C. Williams. "Vanquishing virtue: the impact of medical education." Academic Medicine 76, no. 6 (2001): 598-605.

162. Walker, Lawrence J. "Sex differences in the development of moral reasoning: A critical review." *Child development* (1984): 677-691.

163. Godkin, M. D., K. Faith, R. E. G. Upshur, S. K. MacRae, and C. S. Tracy. "Project examining effectiveness in clinical ethics (PEECE): descriptive analysis of nine clinical ethics services." Journal of Medical Ethics 31, no. 9 (2005): 505-512.

164. Cowley, Christopher. "A new rejection of moral expertise," Medicine, Health Care and Philosophy 8, no. 3 (2005): 273-279.

165. Sekerka, Leslie E. "Organizational ethics education and training: A review of best practices and their application." *International Journal of Training and Development* 13, no. 2 (2009): 77-95.

166. Murphy, Diana E. "The federal sentencing guidelines for organizations: A decade of promoting compliance and ethics." (Iowa L. Rev. 87 2001): 697.

167. Andre, Judith, "Goals of ethics consultation: toward clarity, utility, and fidelity." *Journal of Clinical Ethics* 8 No. 2:193, (1997): 193.

168. Singer, Peter. *Practical ethics*, (Cambridge university press, 2011), 1-16.

169. Bowen, H. "Ethics versus compliance." Real Estate Issues 35, no. 2 (2010): 72-74.

170. Fuchs, Lior, Matthew Anstey, Mengling Feng, Ronen Toledano, Slava Kogan, Michael D. Howell, Peter Clardy, Leo Celi, Daniel Talmor, and Victor Novack. "Quantifying the Mortality Impact of Do-Not-Resuscitate Orders in the ICU*." Critical Care Medicine 45, no. 6 (2017): 1019–27.

171. Gaieski, David F., J. Matthew Edwards, Michael J. Kallan, and Brendan G. Carr, "Benchmarking the incidence and mortality of severe sepsis in the United States," *Critical care medicine* 41, no. 5 (2013): 1167-1174.

172. Weaver, Gary R., and Linda Klebe Trevino. "The role of human resources in ethics/compliance management: A fairness perspective." *Human Resource Management Review*11, no. 1 (2001): 113-134.

173. Raymond, Margaret. "The professionalization of ethics." Fordham Urb. LJ 33 (2005): 153.

174. Weeks, William B., Justin M. Campfield, and L. F. A. C. H. E. Les MacLeod EdD. "The organizational costs of ethical conflicts." *Journal of Healthcare management* 53, no. 1 (2008): 41.

175. Hsieh, H-F., and S. E. Shannon. "Three Approaches to Qualitative Content Analysis." *Qualitative Health Research* 15 no. 9, (2005): 1-22.

176. Hoy, Janet, and Erika Feigenbaum. "Ethics In Community Care Making the Case for Ethics Consults in Community Mental Health Centers." Community mental health journal 41, no. 3 (2005): 235-250.

177. Richmond, Douglas R. "Professional responsibility and the bottom line: The ethics of billing." *S. Ill. ULJ* 20 (1995): 261.

178. Hall, Mark A., and Carl E. Schneider. "The professional ethics of billing and collections." JAMA 300, no. 15 (2008): 1806-1808.

179. Shukla, Ramesh K., John Pestian, and Jan Clement. "A comparative analysis of revenue and cost-management strategies of not-for-profit and for-profit hospitals." Journal of Healthcare Management 42, no. 1 (1997): 117.
180. Tovino, Stacey A. "Hospital Chaplaincy under the HIPAA Privacy Rule: Health Care or Just Visiting the Sick." *Ind. Health L. Rev.* 2 (2005): 51.
181. Light, Donald W. "The practice and ethics of risk-rated health insurance." *Jama* 267, no. 18 (1992): 2503-2508.
182. Wolf, Susan M. "Health care reform and the future of physician ethics." *Hastings Center Report* 24, no. 2 (1994): 28-41.
183. Goldberg, Stephanie, "Ethics of Billing-A Roundtable," ABAJ 77 (1991): 56.
184. Pegoraro, Renzo, and Giovanni Putoto. *Hospital based bioethics: a European perspective.* Padova: PICCIN, 2007. v.
185. Pegoraro, Renzo, and Giovanni Putoto. *Hospital based bioethics: a European perspective.* Padova: PICCIN, 2007. v.
186. Pegoraro, Renzo, and Giovanni Putoto. Hospital based bioethics: a European perspective. Padova: PICCIN, 2007. 4-8.
187. Pegoraro, Renzo, and Giovanni Putoto. *Hospital based bioethics: a European perspective.* Padova: PICCIN, 2007. v-viii.
188. Pegoraro, Renzo, and Giovanni Putoto. "Findings from a European survey on current bioethics training activities in hospitals." Medicine, Health Care and Philosophy 10, no. 1 (2007): 91-96.
189. Pegoraro, Renzo, and Giovanni Putoto. "Findings from a European survey on current bioethics training activities in hospitals." *Medicine, Health Care and Philosophy* 10, no. 1 (2007): 96.
190. Manners, Ian AN. "The normative ethics of the European Union." *International affairs* 84, no. 1 (2008): 45-60.
191. Häyry, Matti. "European values in bioethics: why, what, and how to be used." Theoretical medicine and bioethics 24, no. 3 (2003): 199-214.
192. Fournier, Véronique, Eirini Rari, Reidun Førde, Gerald Neitzke, Renzo Pegoraro, and Ainsley J. Newson. "Clinical ethics consultation in Europe: a comparative and ethical review of the role of patients." Clinical Ethics 4, no. 3 (2009): 131-138.
193. Hurst, Samia A., Stella Reiter-Theil, Arnaud Perrier, Reidun Forde, Anne-Marie Slowther, Renzo Pegoraro, and Marion Danis. "Physicians' access to ethics support services in four European countries." Health Care Analysis 15, no. 4 (2007): 321.
194. Pegoraro, Renzo, and Giovanni Putoto. Hospital based bioethics: a European perspective. Padova: PICCIN, 2007. 22-27.
195. Pegoraro, Renzo, and Giovanni Putoto. Hospital based bioethics: a European perspective. Padova: PICCIN, 2007. 4-7.
196. Pegoraro, Renzo, and Giovanni Putoto. Hospital based bioethics: a European perspective. Padova: PICCIN, 2007. 4-14.
197. Pegoraro, Renzo, and Giovanni Putoto. Hospital based bioethics: a European perspective. Padova: PICCIN, 2007. 22-32.
198. Pegoraro, Renzo, and Giovanni Putoto. Hospital based bioethics: a European perspective. Padova: PICCIN, 2007. 8.
199. Pegoraro, Renzo, and Giovanni Putoto. Hospital based bioethics: a European perspective. Padova: PICCIN, 2007. 10.
200. Fournier, Véronique, Eirini Rari, Reidun Førde, Gerald Neitzke, Renzo Pegoraro, and Ainsley J. Newson, "Clinical ethics consultation in Europe: a comparative and ethical review of the role of patients," Clinical Ethics 4, no. 3 (2009): 131-138.
201. Pegoraro, Renzo, and Giovanni Putoto. Hospital based bioethics: a European perspective. Padova: PICCIN, 2007. 134-138.
202. Pegoraro, Renzo, and Giovanni Putoto. Hospital based bioethics: a European perspective. Padova: PICCIN, 2007. 60-70.

203. McLean, Sheila, and Gerard Magill. First do no harm: law, ethics and healthcare. Aldershot, England: Ashgate, 2008. 110-115.
204. Nielsen, Richard P. "Can ethical character be stimulated and enabled? An action-learning approach to teaching and learning organization ethics." *Research in Ethical Issues in Organizations The Next Phase of Business Ethics: Integrating Psychology and Ethics.* 587-593.
205. Sims, Ronald R. "The institutionalization of organizational ethics." *Journal of Business Ethics* 10, no. 7 (1991). 498-505.
206. Nielsen, Richard P. "Can ethical character be stimulated and enabled? An action-learning approach to teaching and learning organization ethics." *Research in Ethical Issues in Organizations The Next Phase of Business Ethics: Integrating Psychology and Ethics.* 595-600.
207. Carlson, Dawn S., and Pamela L. Perrewe. "Institutionalization of organizational ethics through transformational leadership." Journal of Business Ethics 14, no. 10 (1995). 493-503.
208. O'Kelly, C. "Taking Tough Choices Seriously: Public Administration and Individual Moral Agency." Journal of Public Administration Research and Theory 16, no. 3 (2005). 396-398.
209. Gallagher, John A., and Jerry Goodstein. "Fulfilling Institutional Responsibilities in Health Care: Organizational Ethics and the Role of Mission Discernment." Business Ethics Quarterly 12, no. 4 (2002). 420-433.
210. Gallagher, John A., and Jerry Goodstein. "Fulfilling Institutional Responsibilities in Health Care: Organizational Ethics and the Role of Mission Discernment." Business Ethics Quarterly 12, no. 4 (2002). 435-440.

# Chapter 5
# Concluding Remarks

In many respects, requiring terminal degrees from HEC-C applicants limits the field and growth of ethics as a vocation. Currently, there is a significant need for clinical ethics experts in health care. Requiring advanced degrees and prerequisite requirements for health care ethics consultation certification thins the herd of applicants. Though more stringent criteria for certification may lower the number of certified consultants, those certified consultants will undoubtedly serve as the more formidable ethicists. As the field progresses, it may be the case that the clinical ethics market floods to such an extent that certification standards become more strict. Additionally, growing technologies inevitably raise new moral questions in medicine, and medicine shows no signs of abating when developing new technologies.

The ASBH has made the most significant and seemingly only effort in creating a credential for clinical ethicists. No previous licensure or certificate existed before the HEC-C certification, and various health care professionals have already sought this credential. Still, young professionals can only wonder what other criteria we should seek as a community when evaluating an ethicist. The field should not be exclusionary but should pursue a higher standard for clinical ethics certification. Increasing the criteria and assessment method for new ethicists is a definitive way to ensure the growth and development of the discipline.

Implementing new curriculum standards wherein ethicists demonstrate their analytic moral reasoning skills cannot be accomplished with the points mentioned in this analysis alone. Instead, consider this text as a directional guide where programs and others may intend their efforts. Even Aristotle indicates that a virtue ethics approach to practical decision-making fails due to the theory's inability to practically apply virtues in each moral situation. In a medical situation, it is unreasonable for a practitioner to pause to identify the active virtues, yet the analytical reasoning skills articulated in this analysis, with proper guidance, become an innate feature of a properly trained ethicist.Stepping back and understandingnormative theories of ethics, like virtue theory, guide educators and professionals toward a shared morality or a common goal of moral education. Virtue is an intended goal or standard

J. T. Bertino, *Clinical Ethics for Consultation Practice*, https://doi.org/10.1007/978-3-030-90182-0_5

which ethics consultation curriculums strive to achieve. Though subtly stated, ethics consultation training curriculums that direct their efforts toward a virtuous outcome may yield ethics professionals that are both versed in the philosophical themes associated with their practice and promote their empathy toward involved stakeholders.

The communicative skills of clinical ethicists should not be understated. Many established ethicists versed in theory do not possess the necessary communicative skills to hold effective ethics consultations. Perhaps the only effective method for learning these skills is by practicing clinical ethics in clinical settings. Naturally, working with actual patients contains various difficulties when training ethicists. In controlled environments, ethicists can learn how to both implement moral theories into mock clinical cases. A controlled setting also provides opportunities for ethicists to examine situations in real-time. The pace of hospital medicine is fast-paced and can be daunting to new health care workers. Developing a safe environment wherein, an ethicist may use their analytic moral reasoning skills and practice off-the-cuff responses with mock stakeholders. Though these simulated situations cannot fully replicate proper ethics consults, they can also serve as a venue for experimental consultation methods. Ethics professionals and facilitators alike can use the virtue-driven moral theory articulated in this analysis to develop their own curriculums.

Certifying and credentialing clinical ethics consultants requires a structured curriculum that is malleable with cultural paradigm shifts and possesses the ability to accommodate technological advancements in an ethically supportable manner. However, rather than establishing a definitive lesson plan for clinical ethics consultants, this project emphasizes the critical components within the core purpose of combining knowledge points with skills. In doing so, analytical reasoning for clinical ethics consultants becomes bolstered in a manner that meets accepted professional standards in other fields.The previous lack of accreditation standards diminished clinical ethicists' expertise, credibility, and purpose amongst a community of trained professionals, families, and patients. The ASBH has recognized these issues and has made significant steps to establish a clinical ethics credentialing standard. Still, the requirements and overall rigor of the certification process are insufficient. Recognized standards of clinical professionalism in ethics lie in academic degrees.

Though significant, terminal degrees in philosophy in ethics or related fields do not often establish a firm basis in practical applications in relevant venues like hospitals. In short, the text argues that ethics consultants require an amalgam educational approach that weds institutional practice with analytical education. The chapters that follow chapter one address the identified educational discrepancy by outlining the historical facets surrounding the emergence of clinical ethics, the ethical implications surrounding clinical ethics consultation and genetics and developing technologies, the standards of clinical ethics consultation, the methodologies of ethics consultation, and the moral reasoning involved with ethics facilitation and virtue. The critical lineage of moral theory and clinical ethics traced throughout this analysis demonstrates the tremendous steps clinical ethics has taken over the past several decades and the necessity for more strict standards for ethics consultation certification. The project presents this historical analysis of clinical ethics by tracing a relevant history of moral philosophy and deriving the philosophical origins of autonomy, paternalism, and consent in contemporary American health care. The lineage of these themes demonstrates the foundations of ethical theory to solidify the importance of critical analysis in clinical ethics. Subsequently, we discuss how relevant philosophical literature and ideas have evolved into contemporary methods and theories for ethics consultation.

These facets of the argument expedite with a discussion surrounding clinical consent and emerging genetic technologies.This discussion serves as a stepping stone that aids the progression of ethics consultation training philosophical analyses in clinical ethics and further moves clinical ethics toward a normative discipline. While the history and development of clinical ethics have arrived at a practice that emphasizes facilitation, this project possesses its foundation in a mutual relationship between the knowledge and skills involved in health care ethics and moral philosophy. The importance of turning a curriculum toward moral philosophy as a prerequisite basis for ethical skillsets yields a more substantial standard for professionalizing clinical ethicists.

The origins of clinical ethics, beginning with the philosophy of Hippocrates, Hippocratic influence ancient moral theory has placed upon modern ethical thinkers like Thomas Percival, Richard Cabot, and Chauncey Leake. The adoption of ancient ideas through modern ethicists has historically developed into a pragmatic approach to health care ethics. However, while the practical approach to clinical ethics aided the development toward formalized contemporary ethical standards, these standards require a philosophical system that accommodates evolving medical developments in technology and genetics.

Though only used as a medium to articulate the educational directionof clinical ethics, investigating alternative thinkers' theories regarding developing technologies can greatly influence curriculum development.The dangers of medicine's relationship with technologycreate unique problems that require sound moral judgments. The discussion elaborates this point by explicitly addressing the risks of modern genetic technologies under the philosophical framework of German phenomenologist Martin Heidegger. Though historically noted for his work in ontology, Heidegger's essay *The Question Concerning Technology* aids this project's thesis by illustrating the philosophical effectiveness of using ontological arguments to resolving modern ethical discrepancies. Stressing the importance of a clinical ethics consultant's ability to analytically reason amidst value discrepancies and moral uncertainty is a key skill that this project aspires to articulate.

The clinical ethics consultation methods mentioned in this analysis possess contemporary relevance. A more precise depiction of a revised consultation curriculum emerges by analyzing and amalgamating the beneficial aspects of various clinical consultation methods while simultaneously establishing a philosophical basis for these methods. Defining clinical ethics in a manner that depicts the goals and functions of consultation methods provides more significant insights to educators and students alike. The chapter further explains these points by framing a clinical case around the foremost contemporary clinical consultation approaches, including

Johnson's Four Topics method and Orr and Shelton's Process and format approach. In doing so, the discussion subsequently weds beneficial aspects of existing consultation methods with thematic, philosophical qualities. With the critical aspects of clinical ethics consultation methods extracted and the state of current certification efforts articulated, the discussion migrates to emphasizing the importance of virtue-identification for clinical ethics consultants. Knowledge areas such as informed consent, conflicts of interest, refusal to treat, and medical futility are necessary knowledge components for clinical ethicists. However, ethics consultants must understand these components with an accompanying ability to philosophically deliberate with analytic moral reasoning skills. These skills aid a consultant's ability to mediate conflict and deescalate volatile situations. The project's emphasis on virtue identification and the inherently acquired moral reasoning skills promotes a clinical ethicist's ability to provide constructive recommendations amidst value-laden discrepancies in therapeutic and experimental research.

Each residency or ethics training program inevitably requires modification to accommodate a virtue-based curriculum component for ethicists. Additionally, each residency program that adopts this approach must structure these new elemental aspects of a curriculum around an individual hospital's guidelines and certification standards. The project discusses the practical instances where the proposed curriculum points may manifest, including written exams, portfolios, and bedside charting, all of which are accessible to training programs for other health care professionals. However, difficulties accompany the proposed venues in which the critical curriculum points may become implemented into a certification program. Issues surrounding the cultural acceptance of clinical ethicists in hospital medicine, viewing ethics consultation as a service, and identifying the extent of ethical utility throughout a multitude of health care departments aids the discussion's goal of implementing curriculum components.

The final chapter details Renzo Pegoraro, Giovanni Putoto, and Emma Wray's efforts and setting the EHBP. By examining a relatively successful attempt at developing a hospital-based ethics training program in Europe, the analysis can assess the issues of translating this effort to American bioethics programs. While the European model is effective, especially regarding the shift from standard deontological methods, a hospital-based bioethics program is only as effective as the skill and knowledge point it attempts to teach. This analysis asserts that the virtue curriculum points identified in this discussion apply to any bioethics program to educate clinical ethicists. While Pegoraro, Putoto, and Wray's approach demonstrates an effective means of delivering information to clinical ethicists, their program has the potential to flourish further by implementing elements of virtue identification and subsequent analytic moral reasoning skills.

Despite the difficulties accompanying a residency program for ethicists, this analysis asserts that a residency program for clinical ethicists rooted in analytic moral reasoning and virtue ensures proper educational aspects of clinical ethics. Although issues like cost, resource allocation, employment, cultural acceptance, and continuing education are still present, this project indicates the possibility of a residency program for ethicists rich in moral theory. This type of residency program is rooted in a curriculum that contains necessary knowledge and skill components that an ethicist must possess if he seeks to effectively perform consultations, clinical meetings, and institutional facilitation. Unlike other professional training programs for clinical ethicists, the text presents a unique approach by introducing specific prerequisite moral attitudes that should accompany curriculum composition. If clinical ethicists intend to receive a proper education that aids in their universal acceptance as professionals, ethicists must become moral stewards in their practice. By training clinical ethicists to become ethical stewards with the proposed curriculum components, any existing or future credential program that aims to professionalize clinical ethicists inevitably possesses educational features that ensure the proper development and practice of moral facilitation, ethical leadership, and analytic moral reasoning.

# Index